IEE COMPUTING SERIES 15

Series Editors: Dr. B. Carré
Dr. D. A. H. Jacobs
Professor I. Sommerville

Design & test techniques for VLSI & WSI circuits

Other volumes in this series:

Design & test techniques for VLSI & WSI circuits

Edited by R.E. Massara

Peter Peregrinus Ltd. on behalf of the Institution of Electrical Engineers

Published by: Peter Peregrinus Ltd., London, United Kingdom

© 1989: Peter Peregrinus Ltd.

British Library Cataloguing in Publication Data

Massara, R. E.
    Design and test techniques for VLSI and WSI circuits.
    1. Electronic equipment. Very large scale integrated circuits. Design
    I. Title II. Series
    621,381'73
    ISBN 0-86341-165-7

Printed in England by The Eastern Press, Reading

Contents

Preface

The origins of this book lie in a series of colloquia organized by a number of Professional Group Committees of the Institution of Electrical Engineers, dealing with the general theme of modern VLSI and ASIC technology. Having been responsible for organizing several of these colloquia while serving on the E10 PG committee (Circuit Theory and Design), I was asked by the IEE to review the various contributions as the basis for a book.

The present text is the result of this review. My aim has been to produce a book that will provide an up-to-date view of VLSI and WSI design and test methodologies, combining introductions to the subjects covered with indications of current research directions and results. The material finally incorporated into the book was, in many cases, radically updated and extended to reflect the most recent ideas. My criteria for deciding what to include were that the resulting text should have a coherent theme and development, and that the contributions should have a real tutorial content; where appropriate, they should also give a flavour of current research and development trends.

The book is relevant to undergraduates studying microelectronic systems design, to postgraduate researchers in the area, and to graduate engineers and managers seeking an overview or introduction to the disciplines of semi-custom and full-custom large-scale chip design.

The book is organized into three major parts:

Part 1 deals with **Design methods for custom VLSI chips**. It is sub-divided into three sections dealing with: gate array design; full-custom design tools; and experiences in teaching semi- and full-custom chip design. Section A, *Gate-array technology and design*, provides a detailed introduction to the gate array concept and to gate-array design systems, and gives an account of the time-scales and costs associated with choosing a gate array design route. The contributions emphasize the factors involved in taking the decision to go for a gate array design style. Section B, *CAD tools for full-custom design*, extends the discussion of dedicated gate array computer-aided design packages to the much wider issue of CAD tools for full-custom chip design. The contributors stress the key role of hierarchical design styles in managing chip complexity, and show

how this is reflected in the design of current-generation CAD tools. Examples are given of high-level hardware description languages which feature this hierarchical capability. Section C, *Software tools in teaching ASIC design*, concludes the first part of the book with an indication of current activity in the sphere of educational chip design projects, including the role of the UK Higher Education Electronics CAD (ECAD) Initiative.

Part 2, entitled **Test and fault-tolerance in VLSI design**, picks up, and develops, the theme of design-for-testability, which is cited in several of the contributions to part A as a vital component in any truly comprehensive design methodology for effective ASIC design. Section D, *Design-for-test (DFT) techniques*, introduces the subject in a tutorial style, and then presents a number of novel DFT techniques, including treatments of test in systolic arrays, and in mixed-mode analogue and digital integrated circuits. Design-for-test has developed a rich literature, and many techniques, whilst of research interest, are not notably practical. Section E, *DFT and ATE systems in practice*, takes a look at some of the economic and practical issues affecting the choice and application of DFT and test strategies, and also gives an introduction to ATE systems and their relation to particular ASIC classes. Section F, *Fault-tolerant design for VLSI*, introduces the topic of fault-tolerant design which complements design for test, and describes new work in this area.

Part 3, **Wafer-scale integration**, introduces this topic as a logical extension to the capabilities of the VLSI device, and one that draws on the concept of fault-tolerant design introduced in part 2. There are two contributions in this part of the book; the first provides a tutorial introduction to WSI, while the second describes new work in the practical implementation of WSI devices.

I would like to acknowledge the invaluable help, in preparing this text, of Dr Katie Petty-Saphon and Mr. John St. Aubyn of Peter Peregrinus. I am grateful for the sound advice of the series editors, which has had a significant effect on the final content of the book. I am, in particular, delighted that the authors, all very busy people, were able to produce such a high-quality set of contributions.

R.E. Massara
University of Essex, 1989

PART 1

Design methods for custom VLSI chips

Gate-array technology and design

Chapter 1

An introduction to gate array design

A.D.P. Green

1.1 Introduction

This chapter is concerned with the class of application-specific integrated circuit (ASIC) usually referred to as a *gate array*. Gate arrays have particular attractions for the first-time designer, because the design process is well supported by computer-aided design systems, and the devices can be manufactured economically in relatively low volumes. Gate arrays differ from other types of ASIC in that the active components of the device are manufactured in advance to a predetermined pattern or 'architecture'. The customisation of the gate array, to create an application-specific design, is achieved by connecting the prefabricated active components together using tracks defined in the custom-designed conducting layers. The main roles of a gate array design system are to allow the correct topologies of these final interconnection layers to be defined, and to verify the final result. The number of interconnection layers varies for different architectures. These user-defined interconnections are often implemented in metal, but other materials, such as polysilicon, may be used. For a single interconnect layer, it is common to provide fixed underpasses within the device architecture that can be used as 'tunnels', allowing the paths of the connections to cross each other. Multiple layers of interconnect, where used, are separated by insulation layers containing contact windows that allow the different conduction layers to make contact as required.

Gate array sales contribute to a significant part of the ASIC market. For the ten largest ASIC manufacturers the gate array market was worth $1044·5 million in 1987. (*Electronic Design Automation*, 1988). Gate arrays are particularly popular because of the relative ease and speed with which they can be designed and the relatively low cost of each device in prototype or small-volume quantities. The unit cost of the devices is relatively low because economies of scale reduce the cost of mass producing the basic gate array architecture. Gate arrays are well-suited for simple replacement of random logic and/or programmable logic arrays, and as a result are often used by designers with very little experience of ASIC design. Gate arrays account for the largest number of design starts. During 1987 86% of gate array designs were smaller than 10 000 gates (Masters, 1988).

The design complexity of gate arrays is such that computer aids are required to assist in the design process. In developing computer-aided design tools, a balance has to be achieved between (i) providing all the features, and access to the workings of the system, that an experienced designer might require and (ii) avoiding the creation of a tool that requires specialist knowledge of the design process, and the workings of the system, for even simple designs to be completed.

Gate array CAD systems have tended to develop in one of two ways:

1 as part of a versatile suite of tools that can implement a design in a range of ASIC technologies;
2 as a dedicated gate array design tool.

In case 1, it is common that a wide range of different gate array architectures will be supported. This flexibility often causes the system to be difficult to use, and it is unlikely that the tools provided will be well matched to the gate array architecture. As a result, a large amount of experienced user-intervention may be required to achieve a successful design. Dedicated gate array design tools are usually targeted on a small range of gate array architectures, and are often available from a silicon vendor to support particular gate array product ranges. Such tools are typically implemented on low-power workstations to reduce initial hardware investment, and as a result are usually low-complexity systems. These systems often require minimum user interaction, but their low complexity may prevent them from successfully completing designs using more than, say, 50–60% of the gates available. The limited range of gate array architectures that such systems will support may also be a significant limitation to their use.

The detailed procedure that must be followed to design a gate array is usually different for each CAD system, however the entire design process naturally falls into five stages:

1 Production of the design in a computer-readable form. Depending on the sophistication of the system, this can be achieved by drawing the circuit diagram on the computer screen, digitising an existing circuit diagram, and/or by describing the circuit in a hardware description language.
2 Verification of the correctness of the design by computer simulation.
3 Physical design, or layout, of the gate array, in which the user's circuit is mapped in to the gate array architecture.
4 Post-layout simulation, when the operation of the hardware design produced at the layout stage is verified. At this stage it is possible to perform a more accurate simulation than that of stage 2 because parasitic capacitances and resistances associated with connection lengths can now be accurately predicted using the layout information.
5 Generation of test patterns to verify the final device, and the generation of

mask files which describe the layers that must be fabricated to produce the design.

The entry and simulation of the original design are areas that are usually well supported and seldom cause significant problems during the gate array design process. Similarly the post-layout simulation, final test pattern and mask file generation processes are often crude but adequate in architecture-specific, low-cost, systems. It is the layout stage of the gate array design that most frequently displays the limitations of the system. The physical design of a gate array is usually approached in several stages. In the first stage, the active components – usually MOS devices that make up the design – are mapped on to the existing pattern of devices on the array. A variety of strategies exist to achieve this, but the aim is to choose the mapping so that the devices may be successfully connected together during the routing phase. The next stage is *global* routing of the array, although not all design systems use a global routing phase. This process attempts to allocate approximate paths to each connection required, thus reducing the demands on the final router and increasing the chance that a 100% automatic routing solution will be found. Different strategies exist for global routing. The majority attempt to spread the interconnections evenly over the gate array routing area. The final process is *fine* routing, in which an attempt is made to determine the exact path that should be taken by each connection.

The physical design stage of a gate array may require a significant amount of informed user-interaction for several reasons. The tool may not be sophisticated enough, or may be too general, to take advantage of any specialist routing resources that have been provided as part of the gate array architecture, such as underpasses. It is also possible that the design may contain an unusually large amount of routing, causing certain areas of the array to become excessively congested. This may particularly occur with designs containing large data buses, or logic blocks with a large number of connections. When the physical design stage of the gate array results in failure, it may still be possible for a solution to be obtained with a small amount of manual intervention. This, however, requires knowledge of the gate array architecture and is potentially very time-consuming; there is also no guarantee that a satisfactory solution exists at all.

Gate arrays are available in a wide range of different patterns. The architectural style will depend on the manufacturer, the size of the array and the number of levels of interconnect available. These architectures currently fall into one of three categories (Green, 1987). *Channelled* architectures have the active components of the array laid out in rows (see Figure 1.1) in between which channels are left to allow for routing between the logic blocks. *Cluster* architectures have the active components laid out in rectangular blocks, and the routing is added to the array in the vertical and horizontal channels left between them. *Sea-of-gates* architectures have the active components spread over the array in a tight lattice structure, and no specific channels are allocated for routing. The sea-of-gates architecture is becoming popular for gate arrays containing more than

10 000 gates, but the channelled architectures are generally favoured for smaller gate counts. The distinction between channelled and block architectures can sometimes become rather blurred because some routing channels are usually provided within the rows of active components in a channelled architecture to assist inter-channel connection.

The limited abilities of gate array design systems to use the available routing resources within the device architecture has caused two common approaches to be adopted. The first approach is to design a dedicated placement and routing package that is particularly suited to a small subset of available architectures. This technique is commonly used by silicon vendors who are usually already committed to a particular, limited, architectural range. The second approach is to have a range of sizes of gate arrays, allowing a design that proves to be particularly problematical to be moved on to a larger device with greater routing resources. This solution will probably incur a price penalty, but this may be insignificant relative to the effort needed by an experienced designer to complete the design on the smaller device.

It is important for the first-time designer to understand the various stages of the design process. This process will now be examined in detail.

1.2 Gate array architectures

Gate arrays differ from other application-specific integrated circuits because they are fabricated in two stages. The first stage of the manufacturing process is the fabrication of the active components. These components are fabricated on to the array in a set pattern, which will be referred to as the gate array *architecture*. Within the array there will usually be positions allocated for input or output pads and, in many structures, areas set aside for the interconnections that will be added later. The partially fabricated gate array is commonly referred to as a *masterslice* wafer (Bray, 1985). The fabrication of the masterslice is performed in relatively large volume and can therefore gain from economies of scale. In addition, bulk processing of the masterslice allows tight control to be maintained on the electrical parameters of the devices, enabling repeatable devices to be produced. Once manufactured, the masterslice can be stored until it is to be customised to a designer's particular circuit requirements. The size of a gate array is quoted as the number of equivalent two-input NAND gates that could be fabricated with the active components available on the device.

The active components on the device are often grouped together into small blocks, referred to as *primitives*, to allow commonly used logic blocks, such as NAND gates or D-type latches, to be created. Primitives are designed to allow these logic blocks to be created with the minimum amount of interconnections between the components. A library of interconnection patterns to create various logic functions is usually supplied by the manufacturer. This library, called the *cell library*, will be specific to a particular range of gate array architectures.

The final gate array design will be provided in the form of interconnect patterns that will be overlaid on to the masterslice to produce the required device. In the simplest case, only a single interconnect layer, usually metal, will be defined. When a single interconnect layer is used, the masterslice usually provides fixed underpasses that allow tracks to cross one another without making electrical contact. It is common for such a single interconnect layer also to contain the required power routing connections. When more than one layer of interconnect is to be used, each layer is separated from the others by an insulating layer, containing a user-defined pattern of contact windows to make the necessary connections between the layers and to active components fabricated on the masterslice. The final interconnection layers require considerably less sophisticated equipment to fabricate than the previous masterslice layers, and so the company that fabricates the masterslice may not be the same company that commits the array for a particular application.

Gate array architectures differ between manufacturers and in the number of

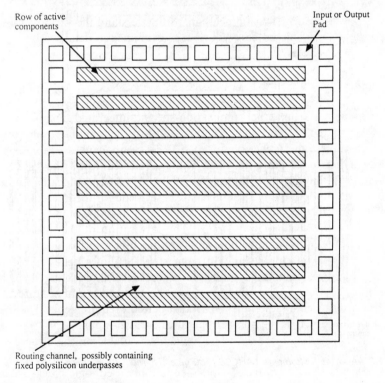

Fig. 1.1 *A channelled gate array.*

interconnect layers required. However, the architectural styles fall loosely into the three categories of *channelled*, *cluster* and *sea-of-gates* structures.

Channelled architectures are commonly used for gate arrays with a small or medium gate count (300–15 000 gates), although larger channelled arrays are used. The channelled architecture consists, as is shown in Figure 1.1, of rows of active components. These rows are usually made up of a repeated pattern of active devices, allowing the logic block, or cell, library to be as simple as possible. In between the rows of active components are areas for the interconnection of the various cells, referred to as *routing channels*. As noted earlier,

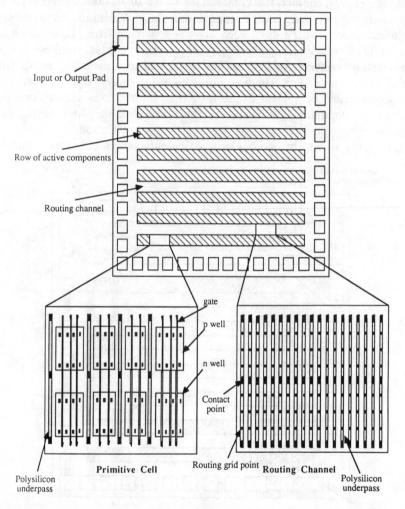

Fig. 1.2 *A Texas Instruments TAHC06 (680 gate) HCMOS gate array.*

architectures that are designed for only one layer of interconnection usually have fixed underpasses fabricated in the routing channel. These underpasses commonly have a standard pattern repeated at the same interval as the active component pattern, simplifying the design process. To allow interconnections between the routing channels, connection paths are often provided that run through the active component rows of the array. To serve as examples, two particular, but very typical, gate array architectures are described here; both are CMOS devices requiring one final layer of interconnection with fixed under-passes contained in the masterslice. These examples are: (i) the Texas Instru-

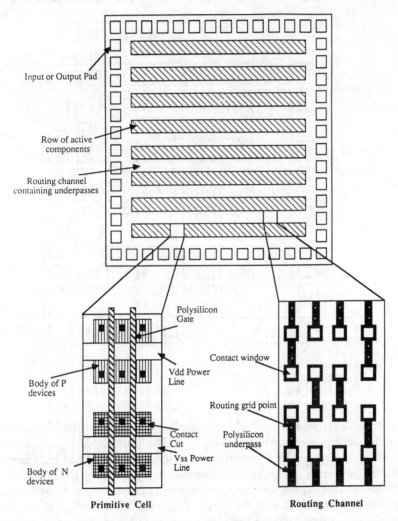

Fig. 1.3 *A Marconi Electronic Devices Ltd. MA8505 (560 gate) CMOS gate array.*

ments TAHC arrays (Texas Instruments, 1983) whose architecture is shown in Figure 1.2 and, (ii) the Marconi Electronic Devices 85 series arrays (Marconi Electronic Devices, 1983a, b) whose architecture is shown in Figure 1.3. The TAHC arrays are available in three sizes: 350, 680 and 1120 gates, and the MEDL arrays are available in 560, 960 and 1440 gate sizes. The MEDL arrays have a repeated pattern of active component cells (see Figure 1.3), usually called *primitives*, each consisting of two p-channel and two n-channel devices. No specific paths are included in the primitive pattern to allow interconnection between the routing channels; should such interconnections be required then contact is made using the polysilicon gates within the primitive cell. The Texas

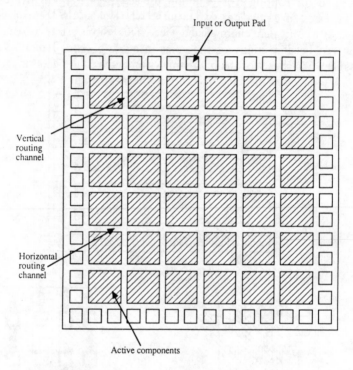

Fig. 1.4 *A cluster gate array.*

arrays have larger primitive cells, each containing ten p-channel devices and ten n-channel devices as shown in Figure 1.2. The primitive cell includes four polysilicon underpasses to allow interconnection between the routing channels.

Cluster gate array architectures (Rangawala, 1982) differ from channelled structures because routing channels are provided in both horizontal and vertical directions over the surface of the array, as shown in Figure 1.4. It is usually much easier to route a cluster array than a channelled array because so much more of the space on the device is allocated to routing. They are particularly useful for designs with unusually large amounts of routing, where greater routing resources will be required to complete the design successfully.

Sea-of-gates architectures (Fujimura, 1988) differ from other gate array architectures in that they have no specific channels or areas allocated for the sole purpose of routing. In this architecture a primitive pattern of active components is copied over the entire array surface in both horizontal and vertical directions, and the only area not containing this pattern will usually be the periphery of the array that contains the input and output pads. Although no particular resources are allocated to routing, it is still necessary to connect logic blocks together. This is achieved by routing between, or over, active areas. As active areas of the array must be used for routing it is common for sea-of-gates architectures, when quoted in terms of equivalent two-input NAND gates, to seem large when compared with equivalent channelled devices. This comparison is often misleading, however, because many of the gates in a sea-of-gates structure may be unrealisable when space for the interconnection between the gates is taken into account. A typical sea-of-gates structure is shown in Figure 1.5.

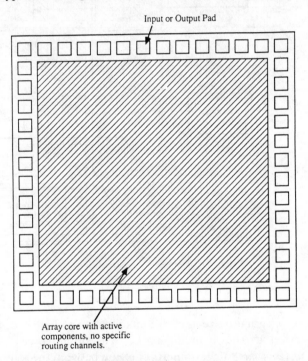

Fig. 1.5 *A sea-of-gates gate array.*

It is difficult when using a gate array to construct complex structures, such as RAM, microprocessors or LSI blocks, without wasting large amounts of the device area. To assist in such designs, some families of gate arrays include, within their architecture, a small number of custom blocks to implement commonly required functions. The remainder of the device will be similar to one of the architectures described earlier.

The architectural style of a gate array will affect the choice of algorithms that will be used to create the physical design for the device. In spite of this, the basic design process for all gate arrays is the same.

1.3 The gate array design process

There are many gate array design systems (Tanaha, 1986, Tien, 1984), and each system will have a slightly different way of achieving the final design. This design will be dependent on the design philosophy adopted and the capabilities of the CAD (computer-aided design) system. In spite of these differences the same basic operations must be performed to complete a gate array design. The entire generalised design process can be represented in the diagrammatic form shown in Figure 1.6.

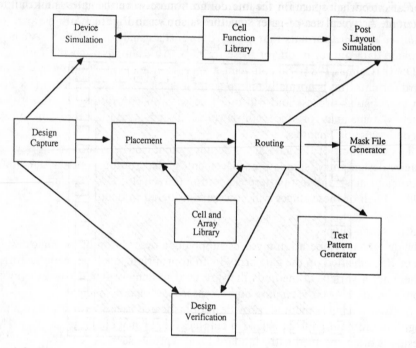

Fig. 1.6   *The gate array design process.*

All CAD systems have three important characteristics: input specification, operation and output specification (Yoshizawa *et al.*, 1982). A gate array design system initially requires a specification of the logic design to be implemented; this design will be provided through whatever design entry system the package provides. In addition to the design, structural information will be provided on the architectures of the gate arrays to be fabricated. Systems usually support the design of a range of different sizes of arrays with similar architectures. However, more general systems may support a range of different architectures. Such architectural data will include information about the functional characteristics of the arrays which can be used during simulation to model the operation of the design as it will be when implemented on the array. The gate array manufacturer will provide a cell library, consisting of approved logic block layouts, and it is usual for the lowest level of the hierarchy in a design to be constructed completely from these library elements. This library information is required by the physical design tools and the simulator. The data libraries that contain the array descriptions and the logical cell descriptions are usually provided by the gate array manufacturers (Texas Instruments, 1983, Marconi Electronic Devices Ltd, 1983b). The availability of such sets of design data files in an appropriate format from the silicon vendors may be a significant factor in deciding which system should be used for a particular design.

The output specification for the design system is very important to allow interface between the designer and the silicon vendor. There are a range of *industry standard* formats that are used and it is usual for a design system to provide data output in one of these formats. An example of such a format is GDS II (Calma, 1984). Silicon-vendor-specific systems may output non-standard formats; this is normally unimportant as such systems are directly targeted on a particular process and will interface with an appropriate manufacturer. Such systems may produce non-portable designs and therefore lack direct second-sourcing facilities.

The operation of the gate array design system can be considered in several stages; design entry, simulation, placement, global routing, post-layout simulation and mask and test pattern generation (Green, 1987, Green and Massara, 1987). Each of these stages will now be considered in detail.

1.3.1 Design entry

The design entry operation involves supplying a description of the logic design to be implemented on the gate array in an appropriate machine-readable form. There are a variety of methods that are used to achieve this; two of the most common are *schematic capture* and *hardware description language.*

Schematic capture methods provide a sophisticated user interface allowing the logic circuit required to be *drawn* in graphical form on a computer display. The design is often specified with standard logic symbols such as NAND gates and NOR gates. Such methods of design entry are particularly helpful to experienced logic designers because this representation of the design is very familiar. The data structure that lies behind the schematic capture package usually consists of a number of linked lists connected by pointers. This type of data structure is used because it is appropriate for connected networks such as circuits, and connections between components and logic blocks within the circuit can be easily represented by pointers between data blocks. Most schematic capture systems will support representations of hierarchy. This is a design method where the circuit can be defined in small sections. Each user-designed section can be stored away and effectively becomes an addition to the cell library. These user-defined blocks can then be used to design blocks of greater complexity. The top level of the hierarchy will be a single block that defines the entire gate array design, and the bottom layers of the hierarchy will consist of *leaf cells*, the cells that were originally provided in the cell library. A diagrammatic representation of a hierarchical gate array design in shown in Figure 1.7. Schematic capture interfaces are generally easy to operate and require little special learning, and the technique lends itself well to implementation in the modern WIMP (windows, icons, menus, pointer) environment. A major disad-

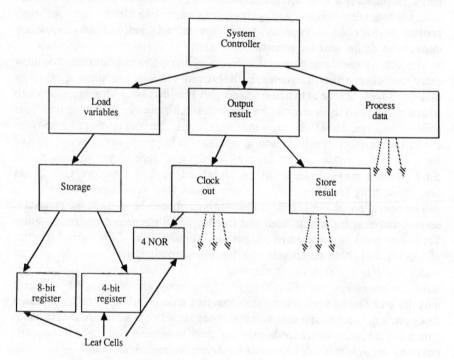

Fig. 1.7 *An example of a hierarchical design.*

vantage of this method of design entry is the speed at which it can be used. Drawing every logic block and connecting these blocks together can, as with the production of an engineering drawing, be very time-consuming, particularly with circuits that are not easy to express in a hierarchical form.

Encoding the design in a hardware description language (HDL) is another major entry technique used by gate array design systems. Hardware description languages are specially developed to aid a designer in describing a circuit. Some of these have been developed from programming languages such as Pascal and Modula-2 with the usual procedure calls replaced with logic functions. A typical HDL definition of a two input NAND gate with inputs IN1, IN2 and output OUT could be as follows:

NAND2(IN1, IN2, OUT);

Other hardware description languages are more developed and contain specific functions and syntax to allow complex hardware structures to be described in a compact and efficient way. HDLs usually produce a circuit description that is less *readable* than the equivalent descriptions produced by a schematic capture package. However, the clarity of an HDL is related to the logical constructs it allows to be used. The experienced designer can use a hardware description language to describe a hierarchical logic design in a compact form, which would take much longer to draw with a schematic capture package. Some systems provide both design entry methods, so that some hierarchical blocks can be *drawn* and others can be encoded in a hardware description language; this arrangement gives the designer maximum flexibility. Systems that provide both entry options usually compile the HDL into the same data format as that produced by the schematic capture package. This is because the data structure will have been designed to be well suited to a circuit description, and generally no penalty will be incurred regardless of the mix of entry techniques chosen.

1.3.2 Simulation

Once the design has been converted into a form that is suitable for use by the gate array design system, it is necessary to verify that the logic description is correct and that the circuit described the function as expected by the designer. This verification is achieved by *simulation*. In order to simulate a circuit, a set of test waveforms to be applied to the inputs and a set of outputs or internal nodes are defined. Test waveforms are often described using a waveform description language (QUDOS, 1985), whereas outputs and nodes to be examined may be described by an interactive command or included in the waveform description file or circuit description file. The majority of commercially available gate array design systems are implemented on hardware that supports sequential program execution. A sequential machine is designed to perform operations one at a time without any true parallel processing. However, a logic system is a parallel machine because each gate of the circuit is constantly evaluating its inputs and, after the gate's propagation delay, providing appropriate outputs. Simulation packages artificially create this parallel environment by working to a virtual timescale. A simulator will usually have a virtual clock and all operations that are performed while the clock is set to a particular value are considered to occur simultaneously. The propagation delays of gates are commonly simulated by using event queues. Any changes to the output of a gate that have been detected are queued for the gate's propagation delay time in the event queue. Before the clock is incremented, the event queue will always be examined for any events that have been queued to occur at this clock time. One of the inaccuracies associated with simulation results occurs because of the simulator's virtual time scale. The time is quantised into discrete periods and the minimum width of these periods is a single *tick* of the simulator's virtual clock. If the design can produce any transient signals that may exist for less than the minimum time step of the simulator, then they will probably be ignored unintentionally by the system. As a result all possible problems that may occur within the circuit may not be shown by the simulator.

The results obtained by using a simulator will always be dependent on the amount of information available to the system about the characteristics of the array and the cell library to be used. Such information is usually provided by the silicon vendor together with the cell library and the array architecture files; examples of these libraries are given in Texas Instruments (1983) and Marconi Electronic Devices Ltd (1983b). The simulator needs to know the propagation delays of each cell in the cell library. This information may be provided as a set of values for each cell defined by the manufacturer or as a description of the delays for each active device on the array, depending on the complexity of the simulator package. The simulator also requires information about the effects that the addition of tracking will have on the array in terms of stray capacitances and resistances. Such information will be particularly important if the simulation information describes the logic cells in terms of nets of active devices, as this information will allow the internal delays within the cells to be calculated. Because gate arrays are manufactured in large volume, using proven fabrication methods, the gate array masterslice is a very consistent product allowing accurate simulation of designs.

The format of the output of the simulator is very system-dependent. Older systems tend to have less elaborate user interfaces and the output will often be columns of numbers representing different logic levels. More recent systems, particularly those implemented on low-cost workstations with colour graphics, often display the simulator output in graphical form. An example of a typical system that fits into this category is the QUDOS Quickchip system (QUDOS, 1985) running on the Acorn Archimedes A440 Workstation.

1.3.3 Placement

The basic architecture of a gate array is fabricated in advance of any design operations, and the position of each active device is already known. *Placement* is the process where each logic block of the design is allocated a position on the array. Each active device in the circuit description is mapped on to an existing device on the array. The process of placement for full-custom and cell-based VLSI designs is relatively unconstrained because active devices are fabricated in appropriate positions as required. However, as the allocated position of the gate array components must correspond to existing features on the masterslice, the constraints on placement for gate array designs are considerable.

The purpose of placement is to allocate the positions of the active devices in such a way that they can be successfully connected together by the routing software. Stray capacitances and resistances caused by long tracking will slow the circuit down and, in certain cases, delay critical signals and cause design failures. The placement package will also, therefore, aim to allocate the active devices so that the total interconnect required is minimised. More advanced packages may allow critical connections such as clock lines and data buses to be specified, and these paths will be reduced in length at a cost of increasing lengths of non-critical connections.

As with many other areas of computer-aided design, there is a trade-off between the closeness of the final solution to a theoretical optimum solution and the time taken to compute such a solution. Times taken can vary between fractions of a second to many hours depending on how good a solution is required. Initial improvements to a random placement can produce significant improvements, but later modifications are likely to have reduced effects. The last small amounts of improvement that can be theoretically possible may be difficult to achieve within a reasonable timescale.

The aim of the placement phase is to generate a placement solution that will be as easy to route as possible given the processing time allowed for solution generation. Practical placement packages may contain a range of enhancements to a basic placement algorithm to improve the final solution. Placement packages often feature the ability to reflect cells so that unnecessary crossing over of connections between cells can be avoided. The ability of the placement package to do this is often dependent on the architecture of the array. Many array architectures are designed so that component patterns exhibit a line of symmetry allowing reflection operations by the placement package which may improve the solution obtained. The cell library will usually be designed to assist the placement package where possible – for example, by including cells that are designed to have a large number of equivalent connection points, making connection to them easier. Cells may be designed so that they have the same connection layout on both top and bottom edges (Marconi Electronic Devices, 1983b), and there may be multiple definitions of cells in the cell library so that the most appropriate version can be selected for a particular location. A detailed review of algorithms used for placement is presented in § 1.4 (*p. 22*).

1.3.4 Global routing

A set of points that are to be electrically equivalent by connection are referred to as a *net*, and any connection between a subset of these points is referred to as a *subnet*. The purpose of routing is to connect points in each net of the design so that the connections required within nets are complete. This may not necessarily involve connection of every point in the net as some may already be electrically equivalent. The position of the points in any particular net will have already been decided by the placement process, although there may be sets of points that are already connected together introducing choices as to where a connection has to be made to complete the net. The global routing process takes place before *fine routing*, which is the operation that determines the final path of each track in the design. *Global routing* aims to decide exactly which points in each net will be connected together and the approximate path that each connection will take. For a gate array architecture with channels, the path of the connection will be established in terms of which channels it will pass through.

For a sea-of-gates architecture, the path of the connection will usually be expressed in terms of the cells it will pass through.

Global routing is important because the fine routing process requires large amounts of processing effort, and thus any reduction in the complexity of the fine routing problem can cause a significant reduction in the amount of processing required by the total routing operation. Approximate paths require less computational effort to calculate because fewer details of the gate array architecture must be considered during the positioning of each connection. A common strategy for global routing is to aim to spread the paths evenly over the surface of the array. This strategy is adopted to reduce the danger of generating *hot spots*, which are areas of the array where a large number of paths need to be routed. Such areas may prevent the router from producing a successful result.

It is common for the same types of algorithms to be used to create a global route of an area as those used for fine routing, and these are described in detail in § 1.5 (*p. 27*). Not all gate array design systems have a global routing stage in the physical design of the device, however, for larger devices, global routing is much more common.

1.3.5 Fine routing

The fine routing stage of the design requires the final paths of all the connections needed to complete the design to be determined. Fine routing often requires a large amount of computational effort and a range of different algorithms exist to perform the task. These algorithms are described in detail in § 1.5. The fine routing problem can be significantly reduced in complexity if the connection paths that will pass through an area of the array have already been selected by performing an initial global routing. As the detailed final path of each connection will be determined during the fine routing stage, the algorithm must take into account any relevant design rules or architectural obstructions. This may lead to a complex routing problem.

It is during this final stage of the physical design of a gate array that the inability to complete the design on a particular gate array size and architecture is often detected. This inability to detect the layout failure may have been caused by an unsatisfactory placement or alternatively by an unrealistic global routing solution. Often, the failure to complete the design is only apparent when the final few percent of connections are being added. This usually occurs because the routing resources have become so congested from the addition of previous connections that no successful paths can be found between the required points.

Most routing algorithms are very sensitive to the order in which the different connections are attempted (Abel, 1972). A variety of different connection ordering strategies exist, normally based on the predicted length of the paths to be completed. The successful completion of the fine routing stage will be very dependent on the complexity of the problem presented to it by the placement and global routing stages. There are some types of designs that may prove particularly difficult to route on a gate array. These include, for example, designs with long multiple connections such as clock or reset lines, designs with a large number of logic cells having a high density of connection pins, and designs containing large data-bus connections.

1.3.6 Post-layout simulation

Post-layout simulation is a very important stage of the design process because it is at this point that the operation of the logic design implemented as a gate array is examined. A simulation of the design after it has been entered into the system only verifies that, in theory, the circuit described will operate correctly ignoring any effects of layout. In practice, connections within logic cells and connections between logic cells will be implemented with a material that will have associated with it capacitance and resistance. Connections within the cells will usually be short and will not significantly affect the design – such effects will often already have been considered in the original simulation using the information supplied by the manufacturer for each cell in the cell library (Texas Instruments, 1983, Marconi Electronic Devices Ltd, 1983b). Connections between the cells will usually be considerably longer and therefore have greater capacitances and resistances associated with them.

The simulator used to produce the post-layout simulation will probably be the same simulator that produced the initial simulation. However, it will have the ability to extract the lengths of the paths between the logic blocks and insert appropriate resistances and capacitances into the circuit to give a more accurate description of the actual circuit implemented on the gate array. It is during the post-layout simulation phase of the design that many time-critical sections of the design may display faults. Examples include signals connected to latches, or gates being delayed and misread. A common problem that occurs is the generation of *glitches*, which are very short duration pulses, usually caused by two signals that were designed to reach the inputs of a logic gate at the same time being separated by the difference between their respective path delays.

Post-layout simulation requires information about the sheet resistances and capacitances of the layers that will be used for the interconnections. Such information is provided by the manufacturer in the form of data files which describe the physical architecture and properties of the gate array and are usually called *architecture files*. The post-layout simulation phase is sometimes used to establish the maximum speed at which a device is likely to run; this is often limited by the signal delays contributed by the longest critical connections in the gate array design.

1.3.7 The generation of masks and test patterns

The generation of mask-making information to allow the final gate array to be manufactured is normally a simple automatic process. If the gate array design system is a specific one linked to a particular manufacturer, then the appropriate files required will probably be automatically generated. If the system used is more general, then a standard format for mask files may have to be agreed with the manufacturer, several such industry standard formats exist (Marconi Electronic Devices Ltd, 1983a).

Test patterns must be generated for the final design; these patterns will be used to test the manufactured device. One of the most common faults in manufactured CMOS logic is the *stuck-at* fault, which occurs when the apparent signal level of a logic gate input or output is stuck at a single value. To detect stuck-at faults the test pattern should cause all, or as many as possible, of the nodes of the circuit to be changed during the test cycle.

The length of a test pattern should be kept as short as possible because a long test time for each device will inevitably increase the cost of the final product. In order to have a good coverage of all the nodes of the circuit, and to keep test patterns to a manageable length, the designer may well include special hardware within the design to allow for adequate testing. There are a variety of different approaches including scan path testing – where registers in the design can be independently set up with the test vectors – and BILBO (Built-In Logic Block Observation), where the circuit is designed so that each logical section of the design can be examined individually to determine if a fault exists.

The amount of support available for automatic generation of test vectors will be dependent on the complexity of the system. Most systems will give a measure of the percentage of internal nodes that are tested for stuck-at faults during a given sequence of test vectors. Some systems will automatically generate test vectors to test sections of combinational logic. Few systems will support the automatic generation of test vectors for fully testing sequential logic designs. With complex sequential logic designs it is usually necessary to include additional hardware within the design to support testing. Techniques for designing-in improved testability, and for automatic test generation, are considered in detail in Part 2.

1.4 Placement techniques

Large integrated circuit designs are usually defined in terms of a number of interconnected functional blocks. The purpose of the placement process is to find a position for each block so that certain criteria are satisfied. The most obvious criterion is to provide a placement that will allow all the defined connections within the design to be routed between the blocks (Goto, 1981). However, this is very difficult to evaluate and therefore it is more usual for other criteria to be used. Placement may aim to optimise one or more criteria including minimum total interconnection between blocks (Lauther, 1980, Tsukujama *et al.*, 1983), minimising the total length of a subset of significant or 'critical' connections or minimising areas of predicted routing congestion (Carter and Breuer, 1983).

The range of different placement solutions will depend on the fabrication technique by which the design is to be implemented. With full-custom or cell-based designs, with no limits to the size of the final device and with at least two layers allowed for interconnection, it will always be possible to create a placement that can be successfully routed (Otten, 1983). The main function of placement, in these cases, is to produce a solution that will not waste silicon area and increase the fabrication cost of the design. When a placement has to be provided for a gate array design, where the area and layout of the device is fixed, no such guarantee of a solution that can be routed can be given. Each gate array will have a limited number of active components that can be used and designs requiring additional components will not fit on to the array. In practice it is usually the lack of space available on the gate array for interconnections between the blocks that makes a placement solution impossible (Ting, 1979).

The placement solution obtained directly affects the complexity of the routing problem that must be successfully completed to produce a complete and valid physical design. A range of different algorithms exist to identify locally optimum solutions to the placement problem (Cooper and Brown, 1968, Khokhani and Patel, 1977, Yoshizawa *et al.*, 1979) within relatively short amounts of processing time. Finding a globally optimum solution is, however, a very complex problem (Breuer, 1972). This results in all placement strategies having an implied trade-off between the quality of solution obtained and the processing time required to obtain it.

Placement algorithms vary significantly in the complexity of their approach and the quality of the solution they are designed to obtain. They can be separated into two types, *initial placement* and *successive improvement*. Initial placement techniques start with the original list of logic blocks that are to be included in the design, and the algorithm allocates a position for each of the blocks according to some set of criteria. Successive improvement techniques start with an existing placement solution and attempt to find ways in which this solution can be improved. As a result of its approach successive improvement, as its name suggests, will often produce a range of solutions as improvement takes place. The most recent solution will usually be the best currently generated.

1.4.1 Initial placement algorithms

1.4.1.1 Min-cut: The aim of the min-cut algorithm is to minimise the total number of connections between the blocks. If an allowance is made for different logic block sizes, this may correspond to minimising the length of the total interconnect between blocks (Green, 1987, Breuer, 1979). The algorithm works by costing each logic block according to the number of connections that will be started if that block is added. This value may be negative, if more tracks are completed than started; zero, if the number of tracks started equals the number of tracks terminated; or positive, if the number of tracks started is greater than the number of tracks terminated (see Figure 1.8). The block with the lowest score is the next to be placed. It is possible to perform this algorithm in one or two dimensions. With a full custom design, a two-dimensional approach will be most suitable, but for channelled architectures with the majority of the routing resources in one direction, a one-dimensional approach may be most suitable. The algorithm requires a *seed* or starting point of at least one originally placed cell before it can be applied and the choice of seed cell can significantly change the final result. A problem with the min-cut approach for large designs with high levels of interconnect is the large number of logic blocks that may have the same score for placement. As a result, the method used to select the cell to be placed from those with the same score will significantly affect the final solution.

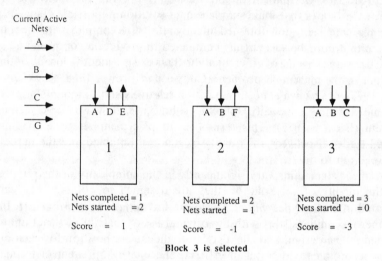

Fig. 1.8 *Selection of logic blocks for placement using a min-cut algorithm.*

1.4.1.2 Grouping/pair-linking: Grouping is a placement technique that attempts to collect together logical blocks that have many connections to each other (Breuer, 1971). These collections will then be placed physically close together on the final device to minimise the interconnection length between the blocks. Pair-linking is a special type of grouping in which each logical block is paired with another with which it has a high connectivity. The process continues by grouping the pairs with each other until the final placement is obtained. Such processes usually give priority to the pairs with the highest amount of interconnectivity. These methods will provide a relatively quick local optimum solution to the placement problem. However, they concentrate on producing a solution based on minimising the total interconnection length between the logic blocks; such a solution will not guarantee that routing can be achieved.

1.4.1.3 Graph/matrix based: There are a range of solutions that approach the placement problem by considering each logic block of the design to be a node of a graph or an element of a matrix and the interconnections to be connecting paths between the nodes (Wing, 1983, Hanan *et al.*, 1977). These methods attempt to minimise the total interconnection length between the blocks, often without considering the difference in the sizes of the logic blocks. Such approaches may yield reasonable results for a two-dimensional problem such as a full-custom design, but the constraints imposed by a gate array design with a limited number of allowed positions for each logic block may make the approach less suitable.

1.4.1.4 Stochastic based: Stochastic placement approaches require the construction of a large number of solutions using random variables to select the different logic block positions. As a result, a large amount of processing effort is required. The usual measure of the quality of each solution is the total length of interconnection between logic blocks, but other measures, such as congestion, could be used. This method is slow at finding good solutions, but it is superior to other, more constrained, approaches because it is not immediately forced to converge on a single local solution. A modification to this approach is to start the search for a solution with a purely stochastic algorithm. After a range of solutions have been generated, the best positions obtained for certain logic blocks are stored and the other blocks are moved. This method reduces the processing effort by initially looking for the best area in which to hunt for a local solution and then concentrating on obtaining such a solution.

1.4.1.5 Simulated annealing: The simulated annealing approach (Nahar *et al.*, 1986) is modelled on the physical processes involved in the cooling of a molten substance, such as glass. During such a cooling process, the stresses between the bonds within the molecules in the material are minimised as a result of the random movement of the molecules. As each molecule moves it will try to find a position that minimises the resultant force on it imposed by the surrounding inter-molecular bonds. The size of movement that the molecule is allowed to make in the hunt for this position is related to the temperature of the substance. This process is simulated by randomly placing the logic blocks on the silicon. A variable referred to as the 'temperature' of the system is set to some large value, and this defines the amount of movement that each logic block can have in searching for an improved position. The connections between the blocks are considered as forces that pull the blocks together, and moves are made with a magnitude related to the temperature of the system and the 'forces' imposed on the blocks by the connections between them. The temperature is gradually reduced to zero, producing a final placement. This method is very powerful because, if the initial starting temperature is high enough, then the chance of selecting a locally optimum solution, rather than one that is globally optimum, is reduced. The quality of the final solution is related to the starting temperature and the speed at which the temperature is reduced. Simulated annealing will produce a high-quality result (Nahar *et al.*, 1986). However, the processing power required is extremely large, and this reduces the possible applications for this approach. It is possible to classify simulated annealing as an initial placement approach because a single solution is provided when the temperature of the system reaches zero. However, the method used to obtain the improvements also allows it to be classified a successive improvement algorithm, because it continually modifies the current placement in an attempt to find a better solution.

1.4.2 Successive improvement algorithms

1.4.2.1 Relaxation/force-directed: This method is used to modify an existing placement (Stevens, 1984). The criterion for improvement is usually the reduction of the total interconnection length between the logic blocks. Each connection that makes up the circuit is considered to be a force drawing together the blocks that it interconnects. The result of the sum of the forces imposed on each block is a resultant force vector with a magnitude and a direction, as shown in Fig. 1.9. The algorithm relaxes the position of the logic blocks having the force

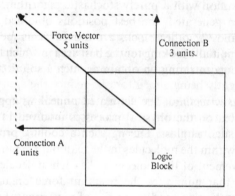

Fig. 1.9 *Force vector placement technique.*

vectors of greatest magnitude and allows them to be moved into unfilled areas of the design that will reduce the resultant force vector. Each logic block is given the opportunity to follow the force vector acting on it by its connections. If the final placement produced after each block has been considered, is better than the starting placement, then the process is continued with the new placement as the starting point; otherwise the process is terminated and the previous starting point is the best that the algorithm can produce. This algorithm tends to draw the logic blocks as close together as possible; and to obtain a sensible solution, especially in full custom designs, it is necessary to leave sufficient routing space to allow the logic blocks to be interconnected during the routing phase. When this process is used to produce a channelled gate array placement, the routing space should be automatically allowed for by the gate array architecture.

1.4.2.2 Pairwise exchange: This approach to successive improvement is very easy to implement because it simply involves swapping two logic blocks in the design (Breuer, 1972). If, as a result, the placement is improved, then the swap remains. Which blocks are to be swapped can be calculated in a number of different ways. The simplest approach is to consider each block with all the others in succession, but this is limiting and can be slow for large designs. Other approaches may involve calculating a force vector for each logic block and giving priority to those blocks with the greatest magnitude vector. This technique is useful because it not only identifies the blocks that could be moved to greatest effect, but it also indicates a preferred direction as shown by the force vector. The computational overheads produced by swapping the blocks can be significantly reduced by simply calculating the difference that will result in the total interconnect length from the block swap; only those connections that make contact with the blocks under consideration need be examined. The pairwise exchange can become more complex in real design cases when the actual logic blocks have significantly different sizes because simple block-for-block exchanges cannot be made easily.

1.4.2.3 Stochastic method: Improving an initial placement with a stochastic search will usually be too time-consuming. However, stochastic methods have the advantage of allowing the possibility of a better local optimum solution to be found than currently being aimed for. To reduce the amount of processing power used, stochastic methods are usually combined with other directed approaches that reduce the number of stochastic operations that are performed before the directed algorithm takes over to find the local optimum placement (Breuer, 1972).

1.5 Routing techniques

Once the placement of the design has been achieved, the routing phase is entered which aims to produce the completed physical design of the integrated circuit. Every design can be defined in terms of a list of logic blocks, or components, from which the circuit is constructed, and a list of interconnections, or nets, that define the appropriate tracks required to implement the design. The general process of routing an integrated circuit with a given placement and a list of nets can be split into four stages; segment generation, layering, path ordering and interconnection generation (Breuer, 1972). The difficulty of the routing task is dependent on the placement solution used, and a usual criterion for judging the quality of the placement is the total length of interconnection between the logic blocks of the circuit (Stevens, 1984). The correlation between this and the difficulty of the routing task is questionable (van Cleemput, 1976). In the case of full-custom or cell-based designs with double interconnection layers, routing can always be achieved by simply expanding the placement to increase the routing resources available on the device, with the penalty of an increase in the total design size (Otten, 1983). With a gate array, where there are only a limited number of positions where logic blocks can be located and the routing resources for the device are predefined, it is possible that no routing solution will exist.

1.5.1 Segment generation

Many of the interconnection algorithms require the task to be split up into segments. Depending on the interconnection method used, each segment will either consist of a single start point and a single destination point or a single start point with several target points with the intention that the algorithm will select the most appropriate target point from those offered during the interconnection process. Segment generation is a simple task for nets that only contain two points to be connected. A general n-point net will be decomposed into $(n - 1)$ segments. Since there will be $n^{(n-2)}$ ways in which this decomposition can be made, care must be taken to ensure that the total interconnection length between these points is not increased by a poor selection of segments. The problem of selecting the optimum segments from an n point net is equivalent to the *travelling salesman* problem which has been shown to be an NP-complete task (Kruskal, 1956). The most common approach is to generate a minimum spanning tree (Loberman and Wienberger, 1957) or a Steiner tree from (Gilbert and Pollak, 1968) the points included in the net.

Consider a simple case. If a net consists of four points to be interconnected in the approximate shape of T as shown in Figure 1.10, then a solution obtained

Fig. 1.10   *An interconnection problem.*

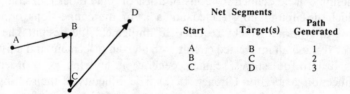

Net Segments		
Start	Target(s)	Path Generated
A	B	1
B	C	2
C	D	3

Fig. 1.11   *A travelling salesman solution.*

using a travelling salesman approach will be as shown in Figure 1.11. If a minimum spanning tree is constructed, then each segment will be a point in the net with its nearest point in the net as the target, producing a solution as in Figure 1.12. Constructing a Steiner tree is more complex than a minimum spanning tree, in fact the problem is NP-complete (Gilbert and Pollak, 1968). Steiner trees often contain additional points, called *Steiner points*, that are added to the net as shown in Figure 1.13. Steiner points need to be added to the net as additional points, and in order to do this the physical position of the Steiner points (or point) included in the net will have to be calculated. This may be a difficult task as their position may create additional constraints on the routing problem.

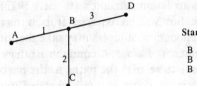

Net Segments		
Start	Target(s)	Path Generated
B	A	1
B	C	2
B	D	3

Fig. 1.12 *A minimum spanning tree solution.*

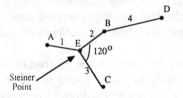

Net Segments		
Start	Target(s)	Path Generated
A	E	1
B	E	2
C	E	3
B	D	4

Fig. 1.13 *A steiner tree solution.*

The process of calculating the appropriate segments may be complicated by the actual arrangement of the device. With a channelled architecture, the paths between the points of the net are often constrained to a limited number of horizontal and vertical channels, and the position of these channels may effect the generation of segments.

An approach sometimes used is to avoid the segment generation process altogether and supply the entire net to the interconnection algorithm. A mini-

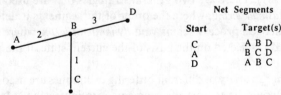

Net Segments		
Start	Target(s)	Path Generated
C	A B D	1
A	B C D	2
D	A B C	3

Fig. 1.14 *A minimum spanning tree generated by routing.*

mum spanning tree is then constructed as a result of the routing algorithm by providing each segment with all of the points in the net that it is not currently connected to as valid targets. This process is shown in Figure 1.14 which illustrates the production of T connections. At the start of the interconnection process, each point will be considered as a subnet with its own individual identifier. The interconnection is complete when all points in the net are labelled with the same subnet identifier. To avoid the generation of unnecessary routing, each connection that is generated becomes a target, allowing 'T' types of connection to be generated.

1.5.2 Layering

In some routing schemes it will be necessary to determine the interconnection layer on which each connection is to be implemented (Breuer, 1972). With a single metal routing, no layer allocation will be required. If fixed, or programmable, vias are available then the allocation of layers may take the form of the allocation of vias to particular segments of a net. A common strategy used for two-layer interconnection schemes is to restrict the paths to the horizontal or vertical directions only, with all horizontal path sections allocated to one layer and all vertical path sections allocated to the other. Many routing systems avoid allocation of particular layers to interconnection paths at this stage of the process and leave such decisions to be taken by the interconnection generation algorithm. Layer allocation can be particularly important in fabrication processes where the interconnection layers have differing electrical characteristics.

In graph theory the problem of layering has been studied by considering the number of colours that are needed to colour a planar graph so that no two similarly coloured sections of the graph will cross. It has been proved that five colours are sufficient to colour all planar graphs, but no graph has been found that cannot be completed with four colours (Ore, 1967). The resultant conjecture, although unproven, is that all planar graphs, and therefore all routing problems, can be completed with four interconnection layers.

1.5.3 Ordering

Net segments are usually routed sequentially, and hence the order in which they should be routed must be determined. Ordering of the segments can have a significant effect on the complexity of the routing problem (Abel, 1972, Heiss, 1968, Hightower, 1981). There are two general strategies that are used to order the net segments: *static ordering*, where the order of the segments is determined before the interconnection process begins, and *dynamic ordering* where the next segment to be routed is selected on the basis of the current state of the problem (Breuer, 1972).

With static routing, a variety of different ordering techniques are used, though the most common would be to sort the segments into increasing 'Manhattan length' (Abel, 1972). The Manhattan length of a segment is the sum of the differences between the horizontal and vertical positions of the start point and the target point as shown in Figure 1.15. With multiple targets, an average is often taken. The segments with the smallest 'Manhattan length' are connected first.

Dynamic ordering usually attempts to give interconnection priority to segments that must pass through the areas of most dense routing on the device. This will have the merit of forcing other segments, having paths available to them which pass through less densely populated areas of the array, to avoid these dense areas. In many cases a global routing operation occurs (Wiesel, 1984) the purpose of which is to allocate the rough path that each segment will take. Static ordering is often most appropriate after global routing as the approximate congestion of the device is known and may be considered in the ordering algorithm.

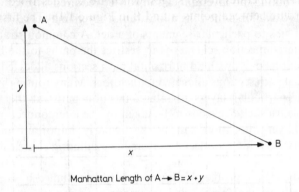

Manhattan Length of A → B = $x + y$

Fig. 1.15 *Calculating the Manhattan length of a path between two points.*

Designs will often contain critical nets which carry important circuit signals such as clocks or reset lines and where significant delays, caused by long interconnection paths, may make the circuit malfunction. Such specified critical nets are often taken into account during the segment ordering process.

1.5.4 Interconnection generation
The interconnection generation process is where the actual path of each segment on the device is established. There are several different types of algorithms that are used for this task: *maze routers* (Lee, 1961), *line routers* (Hightower, 1969), *cellular routers* (Hitchcock, 1969) and *channel routers* (Deutch, 1976). An algorithm used for any interconnection must be able to allow for the avoidance of obstacles and for the inclusion of any design rules that might be imposed by the fabrication process used. Many of these algorithms have been developed for automatic wiring of printed circuit boards and are now applied to the generally more complex task of integrated circuit routing.

1.5.4.1 Maze routers: Maze routers operate on an artificial grid that is considered to cover the area over which the routing will take place. The spacing between the grid points is the minimum allowed path separation. Each grid point can be filled, so that no route may pass through it, or empty; so that later connections may pass through the point. The most common maze router is the Lee router (Lee, 1961), which was developed from previous maze running programs (Moore, 1959). The Lee router has the characteristic of always being able to find a path if one exists, and the path found will be of minimum cost. The basic Lee router will find a path with minimum length, but a generalised cost function may be introduced so that *minimum cost* can take into consideration any required functions such as the use of vias or preferred routing direction (Rubin, 1974). The Lee router is often used as a last-resort router to solve the problems that other algorithms failed to complete (Aramaki *et al.*, 1971, Fisk *et al.*, 1967).

The algorithm operates in two stages, which are now described. We consider providing a route between points A and B in Figure 1.16. The first stage is the

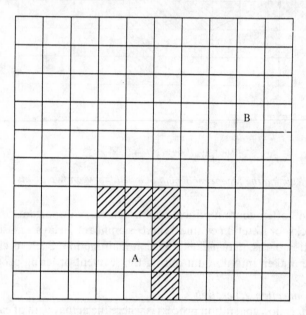

Fig. 1.16   *A Lee routing problem to find the 'lowest cost' path between A and B.*

numbering of the cells. For a cell numbered x all surrounding cells are numbered $x + 1$ if they are not already filled by obstacles or previous routes, and have not already been numbered. This operation is continued until the target cell has been numbered or until no more numbering can be achieved, as in Figure 1.17. If no more numbering is possible and the target has not been numbered then no path exists. The second stage of the algorithm is the trace-back operation where the path is constructed by starting at the target point and tracing the path back through the cells by moving from a cell numbered x to a cell numbered $(x - 1)$ until the start point is reached, as in Figure 1.18.

There are a number of ways in which this basic algorithm can be improved (Rubin, 1974). For example, the number of symbols can be reduced to a

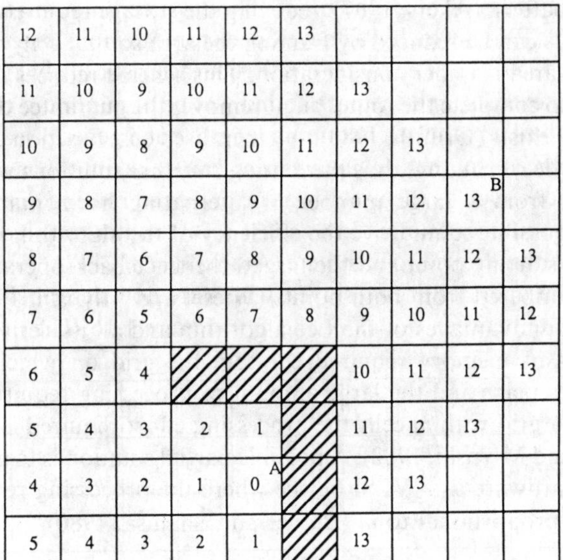

12	11	10	11	12	13				
11	10	9	10	11	12	13			
10	9	8	9	10	11	12	13		
9	8	7	8	9	10	11	12	13 B	
8	7	6	7	8	9	10	11	12	13
7	6	5	6	7	8	9	10	11	12
6	5	4	////	////	////	10	11	12	13
5	4	3	2	1	////	11	12	13	
4	3	2	1	0 A	////	12	13		
5	4	3	2	1	////	13			

Fig. 1.17 *A grid-filling process.*

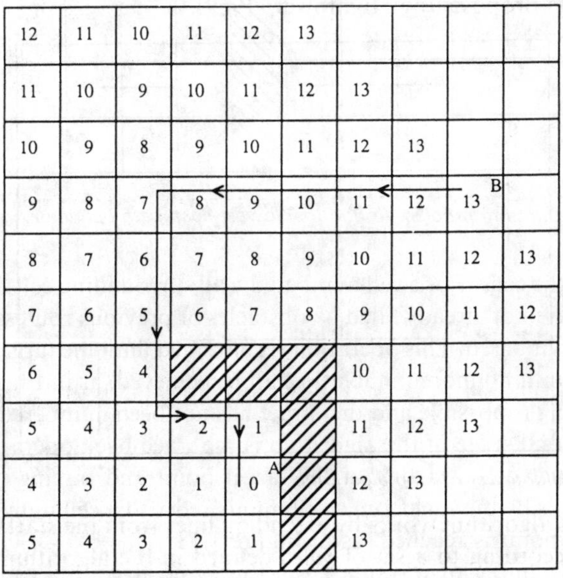

12	11	10	11	12	13				
11	10	9	10	11	12	13			
10	9	8	9	10	11	12	13		
9	8	7	8	9	10	11	12	13 B	
8	7	6	7	8	9	10	11	12	13
7	6	5	6	7	8	9	10	11	12
6	5	4	////	////	////	10	11	12	13
5	4	3	2	1	////	11	12	13	
4	3	2	1	0 A	////	12	13		
5	4	3	2	1	////	13			

Fig. 1.18 *A back trace process.*

1, 1, 2, 2, 1, 1 pattern (Akers, 1967), reducing the storage required. The number of working cells can be reduced by framing the area so that only a small subset of the routing area is under consideration. This scheme reduces the processing time required to complete the route, but it removes the guarantee that a path will be found if one exists (Griffith, 1983). A generalised cost function can be created to effect the style of routing, thus preventing *staircase* solutions where the path is constructed from a large number of alternating horizontal and vertical sections. It is possible to improve the efficiency of the algorithm by numbering from both the starting point and the target, the trace back operating when the same cell is numbered from both points (Wiesel and Mlynski, 1982).

The major disadvantages of the Lee algorithm and all its derivatives are the large amounts of memory required to store the grid on which the routing operation takes place and the large amount of processing required to number each cell. For a grid with m cells the processing effort required is proportional to m^2 (Wiesel and Mlynski, 1982). A possible partial solution is to implement the algorithm is hardware as a VLSI device where the processing required can be reduced to be proportional to m (Breuer and Shamsa, 1980).

1.5.4.2 Line routers: Line routing does not require a large grid of data to be stored. The interconnection path is built up from line segments. Each line segment is stored in a compacted form, typically a vector (x, y, l) where x, y are the coordinates of the lower extreme left point on the line segment and l is the length of the line, with the sign of l indicating a horizontal or vertical segment (Wiesel and Mlynski, 1982). The most well known of these line routing algorithms was proposed by Hightower (1969).

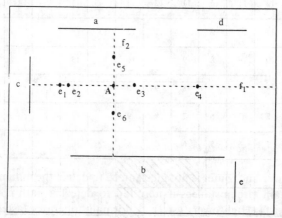

Fig. 1.19 *The Hightower routing algorithm.*

Hightower's algorithm works by extending lines from the start point and the target point according to a set of rules defined in the algorithm. Several new terms that are used in the algorithmic description have to be explained. In Figure 1.19 a point A is shown with three blockages a, b, c, d, e. The algorithm

considers the path possibilities from a particular point called the *object point*. If
A is the current object point then blockages a and b are described as *horizontal
covers* of point A, and blockage c is described as a *vertical cover* of point A. Once
the covers have been identified *escape lines* are generated. A *horizontal escape
line* (f_1) passes through the object point and is bounded by the vertical covers
and/or the bounding rectangle, and a corresponding *vertical escape line* (f_2)
passes through the object point and is bounded by the horizontal covers and/or
the bounding rectangle. On the escape line, *escape points* are identified. On the
horizontal escape line, these are where the horizontal covers finish
(e_1, e_2, e_3, e_4), and on the vertical escape line where the vertical covers finish
(e_5, e_6). The algorithm operates by performing this function from both the start
point and the target point until their line segments cross. As an extension of the
example shown in Figure 1.19, Figure 1.20 shows the same point A with a *target
point* B. Escape lines are extended from B as they were from A, and escape
points e_7 and e_8 are identified. The escape point e_1 is chosen as the new object
point. As the vertical escape line from e_1 is projected, it intersects with the
horizontal escape line projected from point B, and the required path has been
detected as A, e_1, e_7, B.

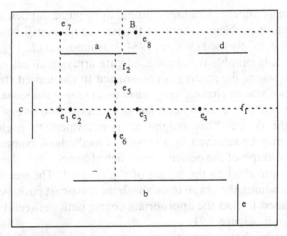

Fig. 1.20 *Path generation with the Hightower router.*

The Hightower line router is not guaranteed to detect the minimum cost path,
in fact in certain circumstances it may fail to detect a path completely even
though one exists (Hightower, 1969). It normally operates faster than a maze
router for any equivalent connection because routing is performed in line
segments and not cell by cell (Wiesel and Mlynski, 1982). The need for large
amounts of memory to be used for grid storage is removed if a line router is
used, however, for many routing problems, particularly where the density of
routing required is high, it will only be practical to use a line router if a Lee

router is also used to complete those connections that are left unrouted by the line router.

1.5.4.3 Cellular routers: The cellular routing method (Hitchcock, 1969) is very similar to the Lee router, in fact the same basic Lee algorithm is used in two stages: the *cellular route* and the *fine route*. Each cell used in the cellular routing phase corresponds to a small cell consisting of a number of the original Lee cells. Routing is performed over this cellular grid, and several connections may be allocated to each cell. This cellular routing process is equivalent to a crude global route. The capacity of each coarse cell is calculated by counting the number of Lee grid points along the appropriate edge. If an equivalent number of connections attempt to cross that edge, then the cell is labelled as *full* and no further connections are allowed. Once the first routing stage is completed, each of the coarse cells is fine-routed using the Lee router and the normal individual Lee cells.

Cellular routing reduces the processing effort and memory required because the size of the grids that must be considered are significantly smaller than for the standard Lee router (Wiesel and Mlynski, 1982). This approach will not easily adapt to irregular structures such as fixed underpasses.

1.5.4.4 Channel routers: Channel routing, as its name implies, is suitable for a range of routing problems where all the net segments will be completed in channels and it is widely used (Wiesel, 1984, Hashimoto *et al.*, 1971, Carter and Breuer, 1983). It is suitable for channelled gate arrays and cell based and full custom designs where the routing is constrained to channelled structures. It is performed in two stages: *channel assignment* and *channel interconnection*. In the channel assignment phase, each net segment is allocated to one of the channels available on the device. This assignment is equivalent to a global routing operation and may be achieved by a variety of methods. A common approach is to construct a graph of the design where each channel is translated into a line with a length equivalent to the length of the channel. The net assignment is achieved by examining the graph to establish the least-cost path. An example of this approach used to find the appropriate coarse path between two points, A and B, is shown in Figure 1.21.

Once each net has been allocated to a number of channels then interconnection can occur. Each channel is considered as a separate entity with the interface between each section clearly defined. Figure 1.22 shows a typical channel-routing problem. The first stage is to construct the horizontal sections that will make up the interconnections (Figure 1.23). Each of these partial tracks must be allocated to a particular row of the channel. This is commonly achieved by using the left-edge algorithm because it will produce an optimal result (Hashimoto *et al.*, 1971). The left-edge algorithm requires each horizontal track to be placed on the first available row, the tracks being ordered so that those that start nearest to the left edge of the channel are placed first (Figure 1.24). The

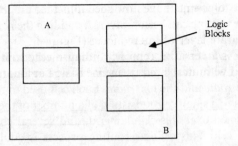

Physical Representation

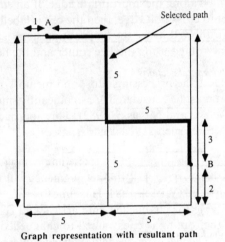

Graph representation with resultant path

Fig. 1.21 *Graph method for channel assignment.*

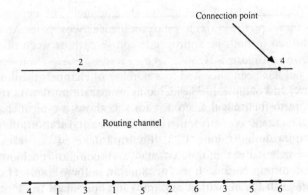

Fig. 1.22 *A channel routing problem.*

horizontal segments of the net have now been allocated an absolute position in the channel and the routing can be completed by adding the required vertical net segments on a separate layer of interconnect (Figure 1.25).

Channel routing is a versatile approach because very complex routing problems can be solved without the requirement of large amounts of processing or

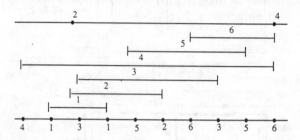

Fig. 1.23 *Generation of horizontal sections.*

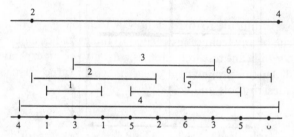

Fig. 1.24 *Application of the Left Edge algorithm.*

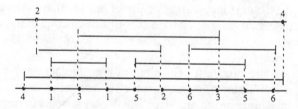

Fig. 1.25 *Construct final solution with vertical track sections in second interconnect layer.*

memory (Wiesel and Mlynski, 1982). Channel routing algorithms can be modified to use a graph-theoretical approach for the allocation of interconnection segments (Sugiayama *et al.*, 1970), this being particularly important for designs that have complex design rules and/or differing track widths. Various modifications can be included in the basic channel routing algorithm to improve the final result. A common modification, for example, is the addition of the facility to include *doglegs* in the horizontal segments to pack segments closer together in the channel (Deutsch, 1976). An example of the use of a dogleg is shown in Figure 1.26.

In spite of the sophistication of many channel routing algorithms, it is not uncommon for a Lee router to be provided to solve particularly difficult interconnection problems that cannot be completed by the channel router. Channel routing is particularly suitable when two layers of interconnect are available. It is, however, rarely suitable for either single-layer routing or single-layer routing with fixed underpasses as these limited resources impose severe restrictions on the operation of the algorithm.

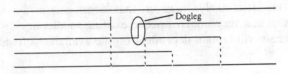

Fig. 1.26 *An example of the use of a dogleg in channel routing.*

1.6 The future of gate arrays

It is always very difficult to predict how a technology will change. This is particularly true of any technology related to integrated circuits because this is an area where significant research effort is currently being expended. It is likely that there will be three significant areas in which large changes will occur; architectures, design tools and cost.

1.6.1 Architectures
Although the great bulk of the existing gate array market is for devices containing less than 10 000 gates it is likely that the average design size will increase significantly over the next few years. The size of market that develops for larger gate arrays will depend on the relative costs of full-custom, standard cell, and gate array manufacturing processes together with the software tools available to support each process.

Current devices are often used for random logic replacement in digital systems. Analogue cell libraries for gate arrays are becoming more common and this will encourage more mixed analogue and digital systems to be implemented as gate array designs.

1.6.2 Design tools
Given the recent advances in personal-workstation technology, an increasing amount of the design process will probably become automated resulting in systems that require significantly less user interaction. Processes such as design verification, correction and the design of test hardware, will become completely automated. Less knowledge of gate array design will be demanded from the designer by the CAD tools and, as these design systems improve, ASIC implementation will become a standard technique. Integrated electronic CAD systems will probably become more common. Such systems will allow a complete

design to be specified and simulated without specifying the final implementation. Once the design has been verified, appropriate sections can be specified to be implemented as an ASIC.

1.6.3 Cost

The cost of designing and manufacturing small gate arrays is already relatively low. As there are fixed costs associated with gate array manufacture, such as capital equipment, such small devices are unlikely to reduce in price dramatically in the future. It is likely, however, that larger gate arrays will probably become available at a very similar price to existing smaller devices. The cost of the design process will remain dependent on the available software tools.

1.7 Conclusions

A few years ago, some market analysts were predicting that gate arrays would become extinct as the cost of cell-based designs reduced. This has proved to be false, and in fact, many of the same analysts now predict that standard-cell-based design approaches themselves are being 'squeezed' between the reducing cost of full-custom design and the increasing complexity and size of new gate arrays.

The increase in CAD support for gate array design, and the relatively low cost of device manufacture, make gate array implementation of both analogue and digital systems increasingly attractive for prototyping and small/medium production volumes.

References

ABEL, L. C. (1972): 'On the ordering of connections for automatic wire routing', *IEEE Trans. on Comp.* pp. 1227–33

AKERS, S. B. (1967): 'A modification of Lee's path algorithm', *IEE Trans on Electron. Comp.*, EC-16, pp. 97–8

ARAMAKI, I., KAWABATA, T. and ARIMOTO, K. (1971): 'Automation of etching-pattern layout', *Commun. Ass. Comput. Mach.*, **14**, pp. 720–30

BRAY, D. (1985): 'Analog/Linear Master Slices' in READ, J. W. (ed.) *Gate arrays: Design and applications.* (Collins, London), pp. 146–83

BREUER, M. A. (1972): *Design Automation of Digital Systems* (Prentice-Hall, Englewood Cliffs, NJ)

BREUER, M. A. (1979): 'Min-cut placement'. *J. Design Automation and Fault Tolerant Computing,* **1**(4), pp. 343–62

BREUER, M. A. and SHAMSA, K. (1980): 'A hardware router', *J. Digital Systems,* **4**, pp. 393–408

CALMA COMPANY (1984): *Stream Format* (17 July)

CARTER, H. W. and BREUER, M. A. (1983): 'Efficient single-layer routing along a line of points', *IEEE Trans. Comp. Aided Design,* **CAD-2**, (4), pp. 259–66

COOPER, J. F. and BROWN, J. A. (1968): 'Automated partitioning and placement in the WRAP system', *Proc. Tech. Program. National Electronics Packaging and Production Conf.* (Jan.),

(Industrial and Scientific Conference Management, California), pp. 85–93

DEUTSCH, D. N. (1976): 'A "dogleg" channel router'. *Proc. 13th Design Automation Conf.* IEEE, New York, pp. 425–33

ELECTRONIC DESIGN AUTOMATION (1988): 3(3), p. 12

FISK, C. J., CASKEY, D. L. and West, L. E. (1967): 'Topographical simulation as an aid to printed circuit design'. *Proc. Share Design Automation Workshop,* 4, pp. 17-1–17-23

FUJIMURA, A. (1988): 'Automating the layout of very large gate arrays'. *VLSI Systems Design,* pp. 22–7

GILBERT, E. N. and POLLAK, H. O. (1968): 'Steiner minimum trees'. *SIAM J. Appl. Math.,* 16(1)

GOTO, S. (1981): 'An efficient algorithm for the two dimensional placement problem in electrical circuit layout'. *IEEE Trans. on Circuits and Systems,* CAS-28, (1), pp. 12–18

GREEN, A. D. P. (1987): 'Application specific integrated circuits – gate arrays', *Proc. Aud. Eng. Soc. Workshop on VLSI Design,* 2/1-5, 82nd AES Convention, Audio Engineering Society, London

GREEN, A. D. P. and MASSARA, R. E. (1987): 'Automatic placement and routing of single metal gate arrays'. *7th Conf. on Custom and Semicustom ICs,* Prodex, London

GRIFFITH, G. L. (1963): 'PCLT: An advanced automatic multilayer printed circuit board tracking program' (TRL/1071, GEC Hirst Research Centre, Wembley)

HANAN, M., WOLFF, P. and AGULE, B. (1977): 'A study of Placement techniques'. *J. Design Automation and Fault Tol. Comp.,* 1, pp. 28–61

HASHIMOTO, A. *et al.* (1971): 'Wire routing by optimizing channel assignment within large apertures'. *Proc. 8th Design Automation Workshop,* IEEE, New York, pp. 50–59

HEISS, S. (1968): 'A path connection algorithm for multi-layer boards'. *Proc. Share/ACM/IEEE Design Automation Workshop,* pp. 6-1–6-14

HIGHTOWER, D. W. (1969): 'A solution to line routing problems on the continuous plane'. *Proc. 6th Design Automation Workshop,* IEEE, New York, pp. 1–24

HIGHTOWER, D. W. (1981): 'The interconnection problem: A tutorial', *Computer,* 7, pp. 8–32

HITCHCOCK, R. D. (1969): 'Cellular wiring and the cellular modelling technique', *Proc. 6th Design Automation Workshop.*

KHOKHANI, K. H. and PATEL, A. M. (1977): 'The chip layout problem: A placement procedure for LSI', *Proc. 14th Design Automation Conf.,* pp. 291–7

KRUSKAL, J. B. (Jr.) (1956): 'On the shortest spanning subtree of a graph and the Travelling salesman problem', *Proc. Am. Math. Soc.,* 7, pp. 48–50

LAUTHER, U. (1980): 'A Min-cut placement algorithm for general cell assemblies based on a graph representation'. *J. Design Automation and Fault Tolerant Computing,* 4(1), pp. 21–34

LEE, C. Y. (1961): 'An algorithm for path connections and its applications', *IRE Trans. Electron. Comput.,* EC-10, pp. 346–65

LOBERMAN, H. and WIENBERGER, A. (1957): 'Formal procedures for connecting terminals with a minimum total wire length', *J. ACM,* 4, pp. 428–37

MARCONI ELECTRONIC DEVICES LTD (1983a): 'Designing on silicon with MEDL' (MDOS 2/5 83)

MARCONI ELECTRONIC DEVICES LTD (1983b): 'System 85 gate array library'

MASTERS, N. (1988): 'The basics of ASICs', *Electronic Design Automation,* 3(3), pp. 15–17

MOORE, E. F. (1959): 'The shortest path through a maze', *Ann. Computation. Lab. Harvard University,* 30, pp. 285–92

NAHAR, S., SAHNI, S. and SHROGOWITZ, E. (1986): 'Simulated annealing and combinatorial optimization', *Design Automation Conference,* IEE, New York, pp. 748–52

ORE, O. (1967): *The four color problem* (Academic Press, New York)

OTTEN, R. M. J. M. (1983): 'VLSI Layout': *Hardware and Software Concepts in VLSI* (Van Nostrand Reinhold, New York)

QUDOS (1985): *Quickchip User Manual* (Qudos Ltd.)

RANGAWALA, Z. (1982): 'Bipolar gate array topology design for auto-placement and auto-routing', *Proc. IEEE* CICC, **1**, pp. 28–31

RUBIN, F. (1974): 'The Lee path connection algorithm', *IEEE Trans. on Comp.*, **C-12**, (9), pp. 907–14

STEVENS, K. R. (1984): 'A review of current placement and routing techniques for integrated circuits'. *IEE Coll. Digest No.* 1984/99 (21 Nov.)

SUGIAYAMA, N., NEMOTO, S., KANI, OHTUKI, T. and WATANABE, H. (1970): 'An integrated circuit layout design program based on a graph-theoretical approach', *ISSCC Digital Tech. Papers*, **13**, pp. 86–7

TANAHA, C. *et al.* (1986): 'An integrated computer aided design system for gate array masterslices: Part 2', *Proc. IEEE 13th Design Automation Conference*, pp. 399–407

TEXAS INSTRUMENTS (1983): 'Texas Instruments Semicustom HCMOS gate arrays TAHC family design manual'.

TIEN, B. N. *et al.* (1984): 'GALA An automatic layout system for high density CMOS gate arrays', *21st Design automation conference*, pp. 657–62

TING, B. S. (1979): 'Routing techniques for gate array', *IEEE Trans. Comp. Aided Design*, **CAD-2**, (4), pp. 301–12

TSUKUJAMA, S., HARADA, I., FUKUI, M. and SHIRAKAWA, I. (1983): 'A new global router for gate array LSI', *IEEE Trans. Comp. Aided Design*, **CAD-2**, (4), pp. 313–21

VAN CLEEMPUT, W. M. (1976): 'Mathematical models for the circuit layout problem', *IEEE Trans. Circuits and Systems*, **CAS-23**, pp. 759–67

WIESEL, M. (1984): 'Loose routing of gate arrays', *Proc. ISCAS*, pp. 444–8

WIESEL, M. and MLYNSKI, D. A. (1982): 'Two-dimensional channel routing and channel intersection problems', *Proc. 19th Design Automation Conf.*, pp. 733–9

WING, O. (1983): 'Interval-graph-based circuit layout', *IEEE Int. Conf. on CAD.*, IEEE, New York

YOSHIZAWA, H. *et al.* (1982): 'A CAD system for gate array automated design', *Proc. 4th CICC.*, pp. 260–2

YOSHIZAWA, H. *et al.* (1979): 'Automatic layout algorithms for master slice LSI', *Proc. ISCAS*, pp. 470–3

Chapter 2

Cost comparisons for gate array prototyping

P.L. Jones

2.1 Introduction

The semicustom approach to ASIC design is of most interest to the systems designer who sees the potential for a rapid and low-risk implementation of ideas in silicon. The choice in semicustom technology between gate arrays and implementations requiring full mask-set fabrication is governed by the balance of a number of crucial engineering factors. These include the total projected development cost, including design and prototyping, the required system architecture, complexity and performance, also the anticipated market demands in respect of volume and lead time.

The low-volume, low-cost semicustom ASIC market is still dominated by the well-established fixed-channel gate array. Its limitations on speed and architectural flexibility do not impair its suitability for PCB system replacement which is still the major application area for gate arrays because of the clear-cut cost-performance advantage they provide.

Certainly, for designs requiring up to 10 000 gates, it makes sense that the prototypes be implemented on a gate array since this route offers the lowest risk, the lowest up-front cost and the shortest turnaround. If the prototype proves satisfactory, well and good. If not, design iteration can be achieved rapidly at low cost. Should the product require improved performance or greater volume, the design can be moved to standard cell relatively easily; only rarely will a move to a compacted optimised full-custom style layout be necessary. This pattern of prototype development leading to rapid entry to the market with options for product improvements has considerable appeal for the electronic equipment manufacturer and is responsible for the emergence of new CAD tools which are aimed at making the gate array and standard cell approaches attractive to a broader spectrum of potential users.

The cost of a prototype run of a gate array design can be greatly reduced by employing a specialist service geared to the rapid turnaround of low-volume parts. A characteristic of such services is the generation of multidesign wafers, which allow sufficient samples for a particular evaluation yet save on silicon usage. There is no point in manufacturing ten wafers each containing 1000 identical designs if the customer requires only 25 samples for test. Ideally what is needed is a means to commit only sufficient silicon area to satisfy a customer's immediate needs, thus saving both time and materials. The lithographic options to produce multidesign wafers are E-beam mask generation or E-beam direct-write, both of which are currently employed by companies offering rapid turn-around services. The choice between direct-write and masks to fabricate gate array designs depends very much on the scale of the prototype operation, and in particular on whether the customer base is sufficiently large to justify the capital investment needed to establish a dedicated E-beam direct-write facility. However, a point of interest is that E-beam direct-write fabrication is now offered for optimised arrays requiring customisation of all layers at costs which only a few years ago were being demanded for prototypes from a single level metal gate-array. Indeed, one might reasonably question how much longer gate arrays will continue to dominate the low-volume, low-cost market sector for ASICs.

In addition to prototyping, the E-beam machine can also be utilised in low-volume production. The objective should be to keep available machine time fully used to generate maximum income either from multidesign prototype wafers or single-design production runs. At two levels an hour, the realisable revenue for single-design production say £1000 per wafer would more than cover the cost of the installation and its on-going maintenance. However, low-volume production using an E-beam machine is only more cost-effective than an optical mask route if the number of parts required does not cost more in E-beam direct-write time than the cost of a mask.

Prototyping for a gate array thus allows a choice between E-beam mask generation, E-beam direct-write, and a totally optical mask route, all of which have their advantages for fast economic routes to silicon implementation. Fast turnaround as an advantage of E-beam direct-write tends to be overstated. It is well known that merchant mask shops can supply tooling for an urgent new design within 24 hours if required. Multidesign masks manufactured either optically or by E-beams reduce implementation costs as follows. If a basic mask cost is £1000, for 20 designs on a mask, the mask contribution to fabrication falls to only £50. This assumes of course that there are 20 designs ready at the time set for wafer commitment. A gate array service geared to rapid prototyping at low volume must set its prices well below those of the mainstream silicon vendors in order to attract new business from customers who would previously have considered an application specific integrated circuit solution too expensive and too risky. The market sector to be gained is at low volume per design. It follows that the value of business to be expected per customer must also be small. To make prototyping attractive for a customer, the price of engineering samples should not be more than about £1000, and there should be no obligation to provide evidence of future volume before orders are accepted. A price of around £1000 is considered reasonable since it is comparable with prototype costs traditionally encountered using assemblies of standard SSI and MSI components. There is little doubt that for many designers of digital systems, the prototype implementation cost is still seen as a significant barrier to use of gate arrays.

Silicon gate array vendors are continually seeking to improve performance of their products by moving to lower lithographic dimensions for the devices, with tighter tolerances on overlay accuracy. This trend, when combined with advances in manufacturing processes directed towards reducing costs in specifically high-volume production, makes increases in prototype costs inevitable. For example, the alignment accuracy requirements for 2 μm CMOS arrays make the use of wafer-steppers essential. Thus, whereas previously for gate array commitment at 3 μm geometries, low-cost lithography can be achieved satisfactorily using multi-design 1:1 projection masks, at 2 μm geometries a stepper reticle at 5 × may accommodate only a single design and may thus in itself contribute up to £1500 to the cost of a prototype implementation. It is thus in gate array prototyping at 2 μm geometries and below that direct-write E-beam may show a significant cost advantage over optical lithography. It is also recognised that E-beams are also the most cost-effective means of generating 5 × stepper reticles. Thus prototyping of ASIC designs relies heavily on E-beam technology to reduce costs either by direct write-on-wafer, or by 5 × reticle pattern generation or by multidesign masks. The optimum lithographic route is based on a trade-off between turnaround time and prototype cost. It will be shown that the cost of the dedicated E-beam direct-write facility and the proposed level of design throughput are the deciding factors.

2.2 The cost of 1:1 projection masks

Although, to some extent, the selling price for a mask is governed by what the market will stand, the increase in the number of merchant mask houses during the past decade has led to intense competition in respect of price, quality and speed of service. For certain applications, an optically generated mask can be obtained for as little as £300, defect counts are below one per inch and delivery can be made within 48 hours. Precise costs depend on the pattern complexity and on the specification of critical defects in respect of size and density.

A mask plate before processing may cost around £100, but the value of the highest quality 1:1 projection mask produced by E-beam may be in the region of £1000, compared with £500 for the same quality produced by optical pattern generation. In prototyping applications an advantage conferred by the more expensive E-beam mask is that it can more easily comprise patterns for many chip designs. Thus, for an E-beam mask comprising 50 different design patterns costing £1000, the mask cost contribution per design is only £20 per layer. However, it must be understood that increasing the number of designs per mask layer will ultimately lead to a need to commit a greater number of wafers to achieve a yield of sufficient prototypes for subsequent assembly and test.

Suppose that for a single-layer metal gate array the cost of the wafer, once processed, is in the region of £100 and that the chip yield is such that a maximum of 10 different designs can be accommodated, if on average 25 samples of each are to be delivered as prototypes. If the E-beam mask costs £1000 and the wafer £100, the cost per design before assembly is £110. Increasing the design count per layer to 50 will require five wafers to be committed to maintain delivery levels and so the cost per design will now be £30. Such economies are however possible only if sufficient numbers of designs are available for commitment at the same time. A rapid turnaround design therefore must carry the full cost contribution of a single mask.

2.3 The cost of stepper reticles

With the need to delineate features less than 3 μm and with increased demand to employ wafers of diameters greater than 125 mm, it becomes impossible to achieve the overlay accuracies necessary to obtain adequate yield by pattern transfer using 1:1 projection aligners. Step and repeat exposure techniques, long established in the mask-making industry, were thus adapted for operation in silicon wafer processing lines by the introduction of the wafer stepper, or direct-step on wafer (DSW) machine. The DSW differs significantly from the mask-making step and repeat cameras in respect of wafer handling, rapid reticle changing, auto-alignment and wide exposure fields.

The stepper reticles may be produced by optical or E-beam pattern generation using the same processing techniques and materials as in 1:1 mask-making. However, whereas for a 1:1 mask, a finite defect density can be tolerated since multiple chip fields are exposed simultaneously, a DSW reticle must be perfect in respect of pattern data and in being free of fatal clear (pinhole) or opaque defects.

Checking a reticle may require more sophisticated equipment than that required for a projection mask, where the accepted technique of optical comparison of one die against another identical die can be achieved relatively simply and at modest cost. However, a DSW reticle may have only one die pattern per plate. Hence the plate must be compared either with a characterised master or against data, in order to verify its acceptability. Only 100% correct DSW plates can be passed.

Correction of reticle defects is normally achieved using high-power lasers to remove extra chrome from opaque defects and focused ion beams to roughen clear defect areas to reduce their optical transmission. Recently introduced mask repair systems claim a capability to correct both clear and opaque defects by ion beam assisted chemical reaction to cause local etching or decomposition as appropriate.

Once generated, checked and corrected, a DSW reticle requires protection. This is achieved by means of a thin transparent membrain (pellicle) held by its rim on a stand-off former about 0·5 cm from the reticle surface. The cost of pellicle fitting depends on the area to be protected and on the type of stepper employed. Some steppers require pellicles on both reticle plate surfaces, others just on the patterned chrome surface. Pellicle protection may add from £75 to £150 per surface to the materials costs of producing the reticle. A competitive price for an average 5 × DSW reticle will be in the range £1000 to £1500. The exact cost will depend on the pattern area, the pattern density and the defect specification. In general, however, a DSW reticle is likely to be more expensive than a 1:1 projection mask plate of the same dimensions. For gate array prototyping, it is desirable to keep the lithography costs to a minimum. With projection alignment, this is simply achieved by including more designs per mask. However, with DSWs, it is not possible to include more than a few different die patterns in a single exposure field.

Blocked reticles, i.e. those which comprise a block of identical die patterns, say in a 4 × 4 or 9 × 9 array, are commonly used to exploit the full exposure field and so increase the throughput of the reduction stepper. The image field size depends on the lens design but typically areas of up to $400 \, mm^2$, 20 mm × 20 mm, can be exposed with patterns containing sub-2μm features, using a 5 × reduction DSW. Clearly a 2 × 2 array could just accommodate four chips 10 mm square, a useful size for a gate array. Each of the chip fields could be the commitment pattern of a different single-layer metal gate array. Hence the reticle cost contribution per design would be reduced by a factor of four compared with a single design reticle. However, since all four die fields would be different, they would have to be checked against data, rather than compared one with another, and this would increase the cost.

The maximum useful area on a 5 × DSW reticle plate is just over 100 mm square, corresponding to about 20 mm square on wafer. Within the exposure field, there may be for example a requirement to place a 2 × 2 array of die patterns, each die pattern on fracturing giving around 200 000 rectangles for exposure by the pattern generator. Each rectangle requires a single flash for exposure on the reticle plate. Such high flash counts are not unusual for say a 10 mm square chip with sub-2 μm features. The combined flash count for the 2 × 2 array would thus be 800 000. If an optical pattern generator were employed, operating at 30 000 flashes per hour, a total of about 27 hours of exposure would be required. In contrast, a raster scan E-beam pattern generator such as a Perkin-Elmer MEBES, working on a 1 μm address grid, could expose the same reticle pattern in less than an hour. Generally, if the total required flash count exceeds 100 000, E-beam rather than optical pattern generation would be used. The high overheads of establishing and maintaining an E-beam facility are reflected in the higher prices charged for E-beam reticles. However, this is normally accepted by the customer as a natural consequence of the increased complexity of the finished reticles and of the potential value in improving the throughput of the DSWs.

Although E-beams are unrivalled in their suitability for 10 × and 5 × reticle pattern generation, they are less suitable for 1 × reticles. Even though the E-beam resolution may be superior, optically produced 1 × reticles have lower defect densities. Also, for quite different reasons, to be discussed in a later section, direct-write on wafer with E-beams has not found widespread acceptance in replacement of optical DSWs, either in prototyping or in low-volume production. It appears therfore that optical lithography still has a key role to play even though device geometries are already becoming closer to the wavelengths employed in optical exposure systems.

With double-layer metal (DLM) gate arrays the contribution of stepper reticles or 1:1 projection masks to prototype costs may be trebled, or more usually, quadrupled. This observation calls into question the entire economics of supplying low-cost engineering samples for evaluation. For a large DLM gate array, say 10 mm square, the DSW reticle cost alone would be in the region of £5000. One possible alternative would be to use a single-level metal (SLM) gate array for lost-cost prototyping and a DLM gate array for production. This concept of course requires both types of array to have identical libraries for logic design and simulation purposes (Jones, 1987).

A further mitigating factor in the high cost of DSW reticles is that with the improved CAD tools now available, giving accurate simulation in design and fault cover analysis for testing, a very high proportion of digital designs on gate arrays will be correct first time and can proceed directly to volume manufacture. Thus, the same reticles used in prototyping may be used in production with a consequent offset of their initial cost. Even so, it must be recognised that at the prototyping stage, reticle costs will have to be met and there is no guarantee that such outlay will be recovered in production.

A recent paper (Wheeler, 1987) suggests that about one third of new gate array designs require a least one iteration. However, the author, from discussions with suppliers in the UK, believes that the proportion of gate arrays which reach successful production as a result of design iteration to be as low as 10%, whereas designs abandoned at the prototype stage may be nearer the quoted 30% level. Hence, in the context of lowering the risk penalty for a gate array implementation it is essential to offer low-cost prototyping. Otherwise, system designers will look towards alternative methods of achieving prototype integration, for example for programmable logic devices (PLDs) and accept losing the advantage of moving directly to a gate array prototype with its consequent smooth transition to a volume part without need for redesign.

The strategy of using low-cost prototyping as a loss leader to secure profitable volume production has also be be considered. It can of course be argued that all operations within a manufacturing group should show a positive return. However, if the prototyping service is linked to the marketing budget, then it would appear simply as a cost necessary to maintain a market share.

Certainly for the sub-2 μm gate array, prototyping would need significant subsidy to be termed 'low-cost', simply because of the outlay required for stepper reticles. Ideally at least 70% of designs will be correct first time and so those reticles could be used in production to compensate for the prototype cost. If the customer vendor interfaces are good and the CAD is accurate, the proportion of first-time successes should increase, thus reducing the overhead of failures at the prototype state. However, there is no doubt that there are difficulties in quantifying the average risk of failure at the prototype stage so that the entry price to the customer remains attractive. There clearly would be no easy formula; it would depend on the customer's experience with the particular type of design.

The principal characteristic of gate array prototyping using DSW lithography is complete compatibility with established fabrication techniques, using standard production equipment. Hence there would be no real need for additional capital expenditure, but the recurrent cost of reticles would be very high. In an alternative approach, using E-beam direct-write, the initial capital outlay is a dominant factor, but the materials costs are very small indeed.

2.4 The cost of E-beam direct-write

The income per prototype design is set by the need to maintain market competitiveness. Therefore, to justify the capital investment required to establish an E-beam direct-write facility, then the predicted utilisation must be high. Low-cost prototyping of gate arrays can only be achieved using the E-beam direct-write approach if there is an assured regular high throughput of new designs.

The cost of setting up direct-write for single-level metal gate arrays is simple to estimate because it is not essential for the E-beam machine to be incorporated in a silicon foundry environment. The simplicity of single-level metal processing allows location either as an adjunct to a mask-making facility or as a stand-alone gate array commitment service. The latter would require clean air services with temperature and humidity control, resist processing, metal etching and sintering facilities, costing somewhere in the region of about £1 000 000 in total. This assumes that the E-beam hardware itself if at the lower end of the cost-performance range, typically a 10 MHz vector-scan gaussian-beam machine. Purchase of a shaped-beam machine could well increase the set-up costs to around £4 000 000. However, even an E-beam machine of modest performance can be expected to write a layer for gate array commitment in less than an hour. The total UK market in new gate array designs is believed to be currently no more than a few hundred annually. Thus, a yearly requirement for 400 prototype designs, at 20 per wafer, could be met by a single E-beam machine in less than a day. Clearly, it is unlikely that the entire annual load will be available simultaneously; nevertheless the implication is that an E-beam installation for direct-write will have spare capacity.

The annual fixed overhead for a stand-alone E-beam prototyping facility must include equipment maintenance, repayment of loan charges and a capital depreciation allowance. Typically, annual equipment maintenance rates are at 10% of capital cost. Interest rates for short term borrowing over three years are about 15% per annum. Depreciation, based on an anticipated five-year eqipment life, is at 20% per annum. The total fixed annual costs can therefore be set out as follows, for an E-beam facility costing £1 000 000 to establish.

	£
Maintenance	100 000
Depreciation	200 000
Loan charges	440 000
Total per annum	740 000

For a stand-along E-beam direct-write facility, the high fixed overheads to be recouped annually are the major determining factors in the lithography contribution to prototyping cost. Management, labour and materials costs are secondary by comparison. Thus, supposing for example, 400 gate array prototyping runs are anticipated annually, the fixed costs of the lithography facility would contribute around £1850 to the cost of each commitment.

It is important to recognise that the equivalent in annual fixed expenditure of an E-beam direct-write facility can purchase around 500 stepper reticles. Thus, if direct-write is the sole income generating activity for the E-beam, then its cost can only be justified if the projected prototype throughput is likely to exceed 500 single-layer or 125 four-layer customisations annually.

Several authors (Bickley and Wallman, 1987, Wheeler, 1987) have attempted to calculate an hourly cost for operating an E-beam machine. The installation cost, covering annual depreciation and maintenance is divided by the hours available after allowance for 'down' time. The result, somewhere between £75 and £110 per hour depends on the detail of the assumptions made. The higher figure is derived from the fixed annual overheads for a £1 000 000 installation, of £740 000, with an availability of 6720 hours per year, assuming that the E-beam is operated 24 hours per day, seven days per week, 50 weeks per year and with 80% uptime. The lower figure neglects loan charges and assumes an availability of 4000 hours per year.

The result of such a calculation is misleading in the assumption that the E-beam is fully utilised. In practice, because of the small size of the prototyping market,an E-beam machine dedicated to direct-write will spend the greatest proportion of the available time idle. Reticle making during the spare time has been suggested as a possible means of increasing utilisation and earnings. However, this would require an unlikely mix of lithographic processing requirement, one appropriate to mask-making, the other to silicon wafer fabrication, both within one facility. It is therefore more realistic to assume that E-beams will continue to be individually dedicated to specific tasks. The operating cost for E-beam direct-write must thus generally be assumed to be related to job throughput, rather than to time per job.

2.5 Scheduling of gate array prototypes

Lowest costings for engineering sample quantities of gate arrays are achieved by scheduling multiple designs on a single wafer exposed by E-beam direct-write or by means of E-beam generated 1:1 projection masks or using reduction wafer-stepper reticles.

Optical pattern generation is ill-suited for multidie reticles because of the high flash counts required. Also, optical composition of multidie masks is not favoured, due to the aggregate costs of reticles. Thus, it is unusual to include more than about 16 designs on a mask plate prepared entirely by optical means. For this, four reticles, each comprising a different 2×2 multidesign array, are stepped to fill each quadrant of the mask in turn.

With E-beam pattern generation, the maximum number of designs per mask plate is decided by the expected mask yield of the gate array base wafers. Thus to reduce the probability of a particular commitment pattern being placed at a defective site recurring on every wafer, there should be a reasonable spread of different chip positions for each design. However, with E-beam mask pattern generation more likely limitations to the total number of different designs per mask are the number available at one time and in the limit, the memory capacity of the E-beam machine itself. Thus, up to 50 patterns per mask would be considered acceptable for 5 mm square chips and 250 useful sites on a 100 mm wafer, giving a very low mask cost contribution to each design. In such a case, to produce sufficient samples of a single design, it may be necessary to print several wafers, which if the mask and process are verified, is a rapid and low-cost operation using an optical aligner.

The scheduling queue necessary to reduce the mask contribution to the cost of a prototype introduces delays into the gate array fabrication cycle. Thus, low-cost prototyping by this technique is normally offered only on 1 month turnaround. If faster turnaround is demanded, the scheduling queue must be bypassed and the full cost of single-design masks must be borne by the customer.

With E-beam direct-write the strategies necessary to satisfy needs in both low-cost and rapid prototyping are different from the multidesign mask or reticle approaches. In low-cost prototyping the aim with direct-write scheduling is to maximise the number of designs per write-cycle, while maintaining an acceptable number of working devices from each design on a single wafer. There is no advantage to include larger numbers of designs per wafer, since the write time to satisfy a particular order will be the same whether the design is written on one or several wafers. However, writing all examples of one design on a single wafer is preferred because of the obvious advantages gained in ease of testing, inspection and assembly. The author's experience with direct-write on a particular type of single-level metal gate array is that even though more than half the chips on the wafer may routinely be expected to pass on probe test, to absolutely guarantee a packaged yield from one design of five chips, at least three times that number should be exposed. The die size in this example is 4 mm square, there are 360 useful sites on a 100 mm wafer, hence around 20 different designs can be safely accommodated.

In rapid prototyping with direct-write, bypassing the schedule queue enables sufficient samples of the required design to be written with only a marginal extra cost for the unshared wafer.

2.6 Cost comparisons – masks, reticles or direct-write

Economic prototype production of gate arrays, either by direct-write or by E-beam masks, will rely increasingly on the use of E-beam technology. The predicted rapid growth in new VLSI parts will necessarily be at low volume, i.e. application specific. Hence E-beams will have an increased role to play in the exploitation of the high complexity and performance available at lower geometries. Significant stimulation of the gate array market will however be necessary before investment in significant numbers of E-beams dedicated to direct-write can be justified.

The relative economics of a stand alone E-beam direct-write facility against E-beam mask procurement depends on demand for utilisation of the service. If the demand is for only 20 designs a month, and the customer is willing to pay only £1000 for engineering samples, investment in E-beam direct-write equipment costing £1 000 000 does not make sense. By way of contrast, the 20 designs on a single mask would cost no more than £1000 from a mask-house with less than one week turnaround from receipt of PG tapes. This assumes the normal mask-house business keeps its E-beam machine running economically throughout the year, whether or not there is a demand for multidesign plates for the low volume market.

It has been estimated that an E-beam installation costing £1 000 000 might have to generate £740 000 of revenue per year to cover loan charges, depreciation and maintenance. The cost contribution of the E-beam to fabrication of a design, assuming a throughput of 20 per month, is therefore £3083. This compares unfavourably with existing commercial services using E-beam generated optical masks where the charges range from only £600 per design. The sacrifice in turnaround to achieve this low cost is to join a fabrication queue with a 30-day minimum lead-time. The E-beam direct-write can clearly not compete in terms of cost per part at the prototype level unless the customer requires seven day turnaround and is therefore willing to pay a premium in excess of £3000 per design.

In order to make a direct comparison of applying E-beam direct-write or E-beam masks in the commitment of prototype volumes on single-level metal gate arrays, the following assumptions have been made.

(1) An E-beam mask cost is constant irrespective of the number of designs included. For this to be achieved the prototype service must take responsibility for merging multiple designs on to a single-mask generation tape.

(2) The gate array wafer cost is constant.

(3) The die size determines the maximum number of designs allowed for adequate yield of each on a single wafer.

(4) The E-beam direct-write cost is constant per wafer.

In the analysis it is assumed that the direct-write facility is kept fully used. If not, then clearly a constant direct-write cost cannot be assumed to apply. However, whether such high demands will ever be found remains open to doubt in a European market, particularly in the case of gate arrays. With the advent of efficient CAD tools for full custom, the viability of the dedicated direct-write machine seems more certain because of the much smaller minimum throughput required to balance the cost of the mask alternative.

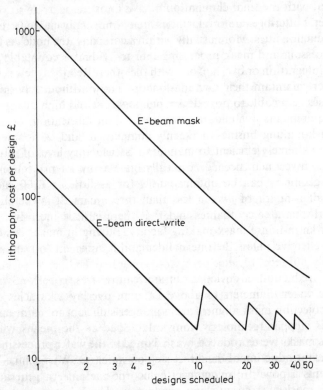

Fig. 2.1 *Prototyping lithography cost for E-beam mask and E-beam direct-write on a single-level metal gate array.*

The decreasing cost per design with increasing design throughput is evident in Figure 2.1 for each of the two methods. However, above a certain limit, determined by chip size and process yield, the direct-write cost stabilises while the mask cost continues to decrease. Although some of the assumptions made may be subject to variation in practice, this observation is believed to be fundamental, illustrating that for direct-write the cost benefit of increased design throughput is not continuous. A high investment in the E-beam installation will raise the plateau of the direct-write commitment cost unless the machine speed and customer base are increased correspondingly.

2.7 Discussion

Multiple designs on E-beam generated masks or reticles can reduce the lithography cost contribution per design only if a number of designs are available simultaneously at the time of fabrication. Alternatively, for individual designs, since no batching is required, direct-write with E-beam can combine rapid turnaround with the total elimination of mask costs.

However, pattern generation facilities are not normally encountered in major wafer production lines. More usually within a company, the policy is to separate wafer processing and mask processing into individual accountable businesses with little integration or overlap, one with the other. Relatively few semiconductor suppliers maintain their own mask-shop. Those without rely on merchant mask houses being able to provide a rapid, low-cost and high-quality service to suit all requirements in standard integrated circuit fabrication.

The mask-making business is keenly competitive and companies involved have to be extremely efficient to maintain a satisfactory level of return on the high capital investment needed. Typically, gate array 1:1 projection masks of moderate geometry can be obtained now for as little as £300 each, a cost reduction of a factor of two in less than three years. It is therefore hardly surprising that wafer companies find that there is little incentive to lay out capital for an in-house mask-making facility. Buying in masks is significantly more cost-effective unless the internal demand is sufficient to keep an inhouse facility fully utilised.

With the reduction of device minimum feature sizes to below 3 μm and the increase of wafer diameters to above 125 mm, overlay accuracies obtainable with 1:1 projection mask aligners are no longer sufficient to maintain adequate yield. Thus stepper technology, long established as the means whereby 1:1 projections masks were produced, was adapted to the wafer processing environment by the introduction of the direct-step on wafer (DSW) machine. However, there remains a powerful reluctance to incorporate pattern generation tools, common in mask-making, in wafer production. The E-beam, used as a direct-write machine, has thus been adopted by specialist units only as a prototyping tool.

The resistance to accepting direct-write E-beam technology in wafer fabrication for other than prototyping therefore requires closer consideration. The E-beam is clearly incompatible in respect of wafer throughput with conventional optical lithography. Whereas a 1:1 projection aligner has a throughput of around 60 wafers per hour, an E-beam can commit no more than about one or two per hour. On the basis of relative throughput, therefore, it can be argued that the lithography cost of an E-beam is 30 times higher than a projection aligner. For prototyping, however, where probably only a single wafer needs to be processed to obtain sufficient numbers of devices for test and evaluation, the dominant factor in the lithography cost is the price of masks or reticles.

For an E-beam direct-write machine and a 5 × DSW of comparable cost, if both are used exclusively for prototyping, the relative economics can be estimated quite simply as follows. The investment, depreciation and maintenance charges will be similar. The DSW produced prototypes carry the cost of a reticle per design, possibly £1000 to £1500, depending on the chip size and complexity, assuming that all commitment layers, up to four for a DLM gate array, will fit on a single reticle plate. If, however, the prototype is satisfactory, then the same reticle can be used for volume production, thus making the reticle cost distribution per part negligible. Consider then moving an E-beam direct-write prototype to production. This will require the manufacture of a reticle, the cost of which will also become negligible in volume production. Clearly the gain in using an E-beam for prototyping is only seen in the reworking of failed designs or in very low-volume production, say of less than 1000 parts, where the reticle cost may still be the most significant item in the total materials and labour involved.

The comparison between E-beam and DSW prototype costs is illustrated by the following example. It is assumed that the reticle cost is £1500, the wafer cost is £100, packaging is at a rate of £3 per device and that only 100 in total are required. The cost of commitment is thus determined as follows, for five wafers:

	£
Packaging	3000·00
Wafers	500·00
Reticle	1500·00
Total per 1000	5000·00
Total per device	5·00
Without reticle	3.50

To the cost of £3·50 without reticle, an estimate of the E-beam direct-write contribution should be added. However, as indicated in earlier sections, the E-beam direct-write cost cannot be estimated from a simple time basis, since the hourly cost of operating an E-beam decreases with increased utilisation. Suppose, nevertheless, that for example an E-beam facility operates with two eight-hour shifts per day, five days per week and 50 weeks per year, the availability would be 4000 hours per year, allowing ample time for maintenance and repair. If the annual cost of establishing and running the facility is £740 000, the hourly cost is £185. Hence at 2 levels per hour, the direct-write cost for five wafers is £462·50, less than one third of the reticle cost in the example.

The question arises however of whether 4000 hours availability could indeed be fully utilised in an E-beam direct-write prototyping facility for gate arrays. Figures quoted by Freer (1987) for CMOS gate array design starts in the UK are 610 for 1987 and 870 predicted for 1988. Add to this a possible 100 bipolar arrays, and the total demand appears to be no more than about 1000 design starts annually. Thus, even if all 1000 designs required four layers of commitment, assuming two levels per hour, a single E-beam would be only 50% used in producing single-wafer prototypes. A more likely situation is that the E-beam is employed in prototyping a much smaller number of designs annually and in their subsequent production at low volume.

Consider therefore the following example which is typical for a large die-size DLM gate array. Assuming 25% yield, 100 die sites per wafer and a requirement for 1000 packaged parts after test, it is necessary to commit 40 wafers. At four levels per wafer, this corresponds to 80 hours of E-beam time, and if the E-beam is fully utilised, a lithography cost of £14 800. If instead, the DSW route had been taken and that all four layers could be placed on a single reticle, the lithography cost would be only £1500. Clearly in this comparison, the E-beam direct-write when used exclusively for low-volume production appears nearly an order of magnitude more expensive than a DSW.

However, in prototyping, the earning potential of a direct-write E-beam is more related to the customer base than wafer throughput. Suppose for example the prototyping market is such that the supplier can afford to assign £1500 to the lithography costs of DLM gate array wafer committal. The £1500 could purchase a single DSW reticle or could contribute to the cost of a direct-write E-beam facility. To justify the latter alternative, the customer base must provide a turnover well in excess of £740 000 annually, which is equivalent in reticle costs to 500 designs. The 500 DLM gate arrays require only 1000 hours of E-beam time, leaving at least 3000 hours available annually for low volume production of 1500 wafers, with the probable expected revenue from each of around £600 bringing in nearly a further £1 000 000. Thus, for an E-beam direct-write facility to be financally viable, it is essential to operate with a high volume of new prototypes and to utilise fully any spare capacity for the production at low volume.

As a final example in this section, consider the lithography costs in E-beam direct-write production of low volumes for gate arrays in general:

Number of parts required $= N_1$

Yield % $\div 100 = Y$

Die sites per wafer $= N_2$

Total wafers $= \dfrac{N_1}{YN_2}$

For each wafer, N_3 levels have to be patterned. E-beam through put $= N_4$ levels per hour. Wafer-write time $= N_3/N_4$.

$$\text{Total commitment time for } N_1 \text{ parts} = \frac{N_1 N_3}{Y N_2 N_4}$$

If the availability in a year is N_5 hours, the total commitment time T expressed as a fraction of a year is

$$T = \frac{N_1 N_3}{Y N_2 N_4 N_5}$$

For an annual running cost of the facility of P, including capital, depreciation and maintenance, the cost C of N_1 parts is given by

$$C = \frac{N_1 N_3 P}{Y N_2 N_4 N_5}$$

For $N_1 = 1000$, $N_2 = 100$, (on 100 mm wafers), $N_3 = 4$, $N_4 = 2$, $N_5 = 4000$, $Y = 0.25$ and $P = £740\,000$, $C = £15\,000$.

This result represents a lithography cost of £15 per chip, which for the large chips implied (i.e. $N_2 = 100$) is not unreasonable. This size die could almost certainly be expected to accommodate designs of up to 10 000 gates in $2\,\mu$m CMOS.

2.8 Conclusions

This article has attempted to explain the role of E-beam technology in low-cost prototyping of integrated circuit designs based on gate arrays. The arguments have therefore concentrated on the fabrication aspects. It is of course recognised that equal attention needs to be focused on the delays which can occur in moving from the committed wafer to packaged chip, including test and assembly. However, for these tasks, most of the lessons have already been learned in the high-volume environment. For example, if a customer wishes a comprehensive evaluation on test he must be prepared to pay. Similarly if there are unusual wiring or package requirements, these also require extra cost.

For gate-arrays in which the devices contain features below $2\,\mu$m and where wafer diameters exceed 100 mm, the choice of lithography tool is between two very expensive types of machine, namely, the direct-write E-beam or the optical wafer stepper. The E-beam suffers from high capital cost, high maintenance, low reliability and low throughput. The wafer stepper on the other hand suffers from high running costs in a prototype environment since the reticles become short lived and very expensive consumables.

Although the single-level metal gate array may seem to be an ideal vehicle for a prototyping service based on direct-write E-beam, the saving mask in cost per design does not in itself justify the capital investment required to establish a stand-alone facility. Thus, a direct-write service will only become viable economically if sufficient customers can be found who will pay a high premium for the rapid turnaround possible with customising a single level of interconnect and the ability to direct-write only the exact number of die required for a given order.

The commercial viability of a prototyping service depends critically on the capital investment and its planned utilisation. Because of the low demand for prototyping, the cost of a stand-alone E-beam direct-write machine is hard to justify. In contrast, E-beam mask and reticle manufacture is a mature process easily adapted to meet prototyping needs economically. Only where there is a requirement for prototyping in sub-2 μm technologies can the E-beam direct-write technique currently compete with optical direct step-on wafer. For the future, a higher performance at lower capital cost and with increased reliability must be achieved for E-beam direct-write hardware to provide economic routes for prototyping and then only if there is sufficient demand for such a service.

References

BICKLEY, J. and WALLMAN, B. (1967): 'Beaming up to fast prototypes without masks'; *Electronic Design Automation,* **21**, (7), pp. 12–13

FREER, W. (1987): 'The economic aspect of fast prototyping in the UK' In SAUCIER, G, READ, E, TRILHE, J. (eds.) *Fast Prototyping of VLSI,* (North-Holland, Amsterdam), pp. 17–24

JONES, P. L. (1987): 'Rapid turnaround for high performance gate array designs' in SAUCIER, G., READ, E., TRILHE, J. (eds.) *Fast Prototyping of VLSI,* (North-Holland, Amsterdam), pp. 157–168

WHEELER, M. J. (1987): 'E-beam direct-write versus reticle stepper technology for ASICs in small volume production', in SAUCIER, G., READ, E., TRILHE, J. (eds.) *Fast Prototyping of VLSI,* (North-Holland, Amsterdam), pp. 195–201

CAD tools for full-custom design

Chapter 3

The application of hierarchical tools in VLSI design and layout

D.W.R. Orton

3.1 Introduction

Electronic system design engineers have at their disposal many means with which to exploit the benefits offered by silicon technology. At one extreme the microprocessor can be programmed to provide the function required, with an ease of modification unmatched by any other system implementation. Gate arrays are more difficult to modify but may offer some benefit over the micro-processor in terms of performance and perhaps ease of interfacing. Standard cell layouts move further from ease of modification but give even higher perfor-mance, while the ultimate performance obtainable with the currently available technology can be obtained by the use of a custom silicon design, where the whole design detail is closely matched to the system requirements. Figure 3.1 demonstrates the range of choice offered to the systems designer. The advan-tages which custom VLSI technology offers may sometimes be matched by the difficulty of achieving them; the initial design and any modifications are costly and may involve relatively long timescales compared to the development of microprocessor software. Examples of the benefits possible are increases in data throughput as high as over 100 times that achievable with a microprocessor using the same technology and major savings in board area and power dissipa-tion. Techniques are constantly being sought to tap these benefits without incurring the great costs traditionally involved.

VLSI designs do not have a monopoly of difficulty. Many system design problems are equally complex and difficult to implement and test. The problems are more exposed in VLSI designs because of the long timescales and increased cost of change should an error be found. The main problem is the issue of complexity and the means of managing that complexity. The accepted method of tackling the problem is by the successive partition of the design information into sections whose complexity can easily be handled in terms of both function and implementation. In a VLSI design many different forms of design data are necessary and the partitioning must attempt to be uniform across the different data representations. Design tools must be employed which support this design strategy and assist the designer in the task of managing the design.

Custom VLSI design is nearly always justified by the need for high performance, and therefore the design focus is performance optimisation. This leads to the need for well-characterised models of the individual elements which will be used in the design. To some extent the VLSI designer has to grapple with problems at all levels of design abstraction from the system down to basic silicon processing problems. This then is the VLSI design environment, from which aspect the problems of VLSI design and the employment of hierarchical design tools will be reviewed.

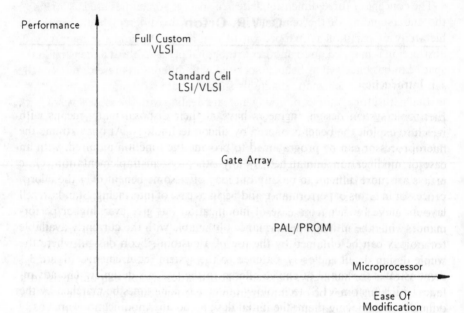

Fig. 3.1 *Technology choice.*

3.2 Design space

A system implemented in VLSI form can be described in many ways. The system designer may view it as a collection of process algorithms, the VLSI applications engineer may consider it to be a set of processors connected together, while the silicon circuit designer could describe the whole as a series of transistors connected together. Other views are equally valid and are relevant to the particular expertise which is being applied to the design at the time. Such representations form a hierarchy where the relationship between one view of abstraction and another can be defined.

It is also possible to consider a hierarchy of partition. If the whole system is regarded as the highest level, it is possible to subdivide the chip into modules each of which perform a specific function. That function may also occupy an identifiable area of the chip layout. Within each module further subdivision should be possible until a primary unit of the type under consideration is reached. This partition can be applied to most design representations to some extent. It should also be possible to relate the functional partition to some equivalent partition in the layout. Such a partitioning is not inevitable; many circuits which are designed place equal significance on all components; the design is said to be 'flat'. It is widely understood now that the partitioning of a design can bring major benefits to the accuracy, integrity and efficacy of the system especially when it is a VLSI device.

The concept of three-dimensional design space is sometimes used as an aid to the understanding of the relationships between the different hierarchies. The hierarchy of partition is mapped on to one dimension and the hierarchy of abstraction is mapped on to another orthogonal plane. A third dimension in the space can be conceived as being associated with different representations of the same abstraction; schematic diagrams showing gate structures compared to a textual netlist for example, both of which are dealing with the same level of both partition and abstraction. Figure 3.2 shows two dimensions in design space representing the hierarchies of partition and abstraction and some examples of design entities in their respective positions. The aim of the design team is to develop all abstractions with as high a degree of equivalence in partition as is reasonable. If this objective is achieved, it should be possible to some degree to verify that the overall function of the design will be realised by the mask layout when the device is manufactured and that the test data will truly test the manufactured part. This ideal can be reached by successive comparison and correlation of the various design representations, provided there is sufficient commonality. It should be noted that not all abstractions need be present at every level in the hierarchy; a system algorithm description can only exist at a few partitioning levels before the partitioning becomes difficult or irrelevant. An example of the abstraction hierarchy could be:

- function (complete chip description in algorithmic terms)
- flowchart of the algorithm
- arithmetic (description of the procedures in integer term)
- logic (description in terms of gates, register elements etc.)
- circuits (transistor realisation of gates)
- mask data (polygon or E-beam raster data).

The partition hierarchy could comprise in part:

- chip
- processor
- arithmetic unit
- bit slice arithmetic unit
- adder
- sum and carry generators
- gates.

It may be difficult in some cases to separate the two hierarchies. At high levels the functional partition may be an essential part of the description. The design approach merely formalises and extends what is to some extent sensible thinking. The designer usually selects the appropriate abstraction for the level in the partitioning hierarchy which is under consideration. It would be difficult, for example, to describe a single gate in algorithmic terms. It is important, however, to be able to compare different abstractions and this is why both hierarchies must be considered.

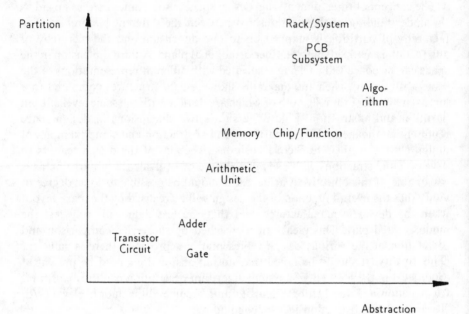

Fig. 3.2 *Design Space.*

3.2.1 Top-down design
In the development of simple systems using breadboard techniques it was perfectly feasible and arguably efficient to design bottom-up, where small modules were developed, assembled, tested and then extended until a complete system was created. In using this technique the overall system requirement is probably never formally expressed; it is held in the designer's mind and gradually takes shape over the period of development. The specification and design of complex systems is essentially an iterative process whereby ideas are explored, modified or discarded until an appropriate choice is found. It is therefore important, especially where design teams are employed, to define clearly the objective and gradually to add detail until a complete implementation has been created. The result of failure to follow this advice will be at least extended development times and at worst chaos. In this top-down method, the abstrac-

tion hierarchy is traversed, with the aim of minimising the length of the iteration loops within that hierarchy. At all times the system designers must check that the current representation meets the implications of the requirement specification. Failure to observe this constraint can lead to large amounts of work being repeated later when errors are detected. The designers must also ensure that the abstraction is developed so that the partitioning is suitable at lower levels of abstraction, particularly layout.

3.2.2 Isomorphic hierarchies

The structure of the hierarchy which is employed in a VLSI design is very important. If comparison and correlation techniques are to be used efficiently to verify the design, or the design aids are to be of reasonable simplicity, it is important that the hierarchy of partition in the functional abstractions and that in the layout abstraction are equivalent or 'isomorphic'. The development of layout is usually based on a library of functional blocks or cells. Typical examples of these cells would be RAM, ROM, registers, arithmetic units, multiplexers etc. In the layout system some method of assembling such modules will probably exist which enables the construction of multiple bit versions of each. Such modules must be created with appropriate data width, then be placed and connected to their neighbours to construct the complete machine. It is unlikely that each bit within a module will be identical. The least significant bit and the most significant bit of an adder module, for example, could well have different structure to those in between due to the different carry requirements. This restriction on bit-level modularity is often ignored in functional terms. Sometimes a bit-slice view of a system is generated where each bit is assumed to be identical. This may be very difficult to produce efficiently in layout terms. It is beneficial if the functional definition of the whole process reflects the physical partitioning, providing each functional module as a macro-function, and not attempting to build the multiple bit process from a bit slice description of the whole. Test requirements may also impinge on the system partitioning and this must also be considered; a 64 bit arithmetic unit or counter could take an impossibly long time to test without sub-partition. It may also be important to be able to be flexible in terms of the physical construction. A rigid library of modules may be extremely difficult to match to the required partitioning for a particular system architecture. For example, RAM and ROM modules with and without the address decoding may enable an address decoder to be shared between a number of modules.

The adoption of this design method can be seen to constrain the system development such that the partitioning of the system must be directed towards the partitioning of the layout. This therefore requires the system designer at least to understand the structure and partitioning applied to the layout library unless a new library is to be built every time a new design is implemented or sub-optimal results are to be obtained. It also places constraints on the layout library developer. Each cell must be capable of connection to an adjacent cell with the minimum of difficulty. These design directives are to some extent independent of the design aids which will be used to implement them; but if the design tools do not readily encompass such a philosophy, then the designer will have great difficulty in producing efficient results. The application of this philosophy to the whole of the design process has been found to make the real difference in performance of the complete system. The design tools must therefore support the use of this design approach; hierarchy, partition, cross-correlation and transfer of information from one medium and abstraction to another must be facilitated. ELLA and ISIS are two CAD systems which to some extent provide the necessary facilities. Figure 3.3 shows how these systems, described elsewhere in this text, can be used to provide a design route from system specification through to layout and test vector definition.

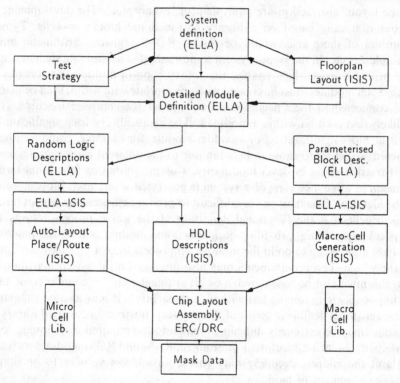

Fig. 3.3 *Design Route.*

3.3 ELLA

ELLA is described in more detail in Chapter 5; the philosophy of use will be principally considered here. ELLA enables a system to be described in terms of integer functions, processes and down to boolean gates and logic function. The design representation may be any mix of these levels of abstraction, provided the necessary interfaces are installed. ELLA is probably not the most appropriate means of description of systems at the highest level of abstraction. At these levels structure is not relevant, and some notion of structure and therefore implementation is inherent in the ELLA description. In order to study the requirement specification of a system in an implementation independent manner it is sometimes necessary to employ conventional sequential programming languages and hardware emulation. This is usually very difficult to relate to the hardware implementation and is only used as a means to establish the system requirement more clearly.

A system design development in ELLA will usually commence with the definition of the system outline, signal naming and types, closely followed by the same operation on the first few levels of partition. Before many levels of partition have been traversed it should be possible, dependent upon the complexity of the system, to provide code in ELLA which can be animated and the results observed. Because the detail of logic design and bit-level communication is absent at this stage, the simulation should be efficient in terms of computer utilisation and response so that the designers can iterate and optimise the design rapidly. In addition, because the system has been broken down into more manageable sections, the designer may be able to consider the implications of the design strategy in terms of the lower levels of hierarchy, both functional and physical. For example, some knowledge of achievable data rates may be available or the width of a communicating channel may be seen to be unrealistic where a partition involves long connection paths in the layout. As the design is shown to be satisfactory at one level of abstraction, the modules concerned can be expanded in detail and the new representation shown to be logically equivalent to the higher-level representation. Such design validation can be carried out in isolation from the whole to minimise the amount of simulation time and the difficulty of data handling. Indeed, where a team structure exists it is important to make sure that only valid models are inserted into the complete design database where others may make use of them.

The representation of functions such as arithmetic and storage will usually be fairly easy in the high-level form described above. It has been found that system modules which contain complex sequential processes are difficult to abstract to a simpler form. The adoption of a simple numerical or named coding of states and the use of 'case' statements in the ELLA code is the best that can normally be achieved. At lower levels of hierarchy, specific state assignment and logic minimisation can be applied to lead to a minimised structure such as a ROM or PLA. Some special purpose tools have been written to enable the generation of these processes from state flow charts in a more schematic form. These tools are currently outside the ELLA environment but can be used to generate ELLA code for animation and integration.

The management of complexity in ELLA can be aided by the use of the context facility to isolate units within the partitioning hierarchy. Common units may be held in a global context from where they can be imported into the individual units. This establishes a commonality of design and avoids duplication of effort. Standards for types of signal, element-naming and common design features such as clock-timing and active levels of signals must also be applied. In all these activities the relationship between the functional and the physical must be given careful consideration. Copious comments are almost essential within the ELLA code in order to describe status and change history of the text. The detail of the design verification must also be agreed between the members of a design team. Low-level modules may be verified in an exhaustive manner; for more complex modules the boundary states of operation may need to be identified and explored. It is frequently found during the verification of the ELLA code that the original specification is incomplete in some way, due to some misunderstanding of the operation of the system or perhaps a restriction of the chosen implementation. This is evidently an extremely important part of the design process and stresses the need for full design validation from the highest level of abstraction down. As well as verifying the modules, communication between them must be checked out under limiting circumstances.

The ELLA environment should enable the designer to gradually add detail to the modules, establishing the partitioning, structure and communication requirements which will apply in the layout. ELLA does not generally provide the designer with information on performance of parametric data. It is perfectly possible to design an entity in ELLA which cannot be implemented with sufficient operational speed or be accommodated in the available chip area. This is the realm of the other design medium under discussion, ISIS.

3.4 ISIS

The ISIS system enables the generation, simulation and verification of descriptions of CMOS and NMOS circuit layout and connectivity. It is based around a structured hierarchy of modules, each of which has associated input and output signals and which can include lower-level modules in a partitioning hierarchy. It supports four principal design abstractions:

- functional modules
- logic modules
- circuit detail
- mask representation.

By the imposition of a system of isomorphic hierarchy whereby each representation is partitioned in exactly similar manner, representations in different abstractions can be compared and verified. The logic and functional module description can be entered in a schematic representation, but is normally generated and held in the textual form which it is possible to simulate. Circuit level simulations can be carried out using a full voltage and current transient analysis or by using a switch level model in which logic levels are transferred with nominated delay. Mask data is normally entered and edited at a level just above full mask data abstraction.

The user interface is a workstation with textual editing and display facilities and a graphics editor and mouse. The latter is used for the schematic and mask representation operations. Transistors and connections are entered symbolically in stick form, the detail being automatically added by the software. Automatic and immediate checks are carried out on the layout in order to ensure that design rules are not infringed.

The application of the ISIS system to the design of a VLSI circuit under the proposed design methodology is in parallel to the use of ELLA. As soon as a structural partition appears in ELLA, an equivalent structure can be postulated

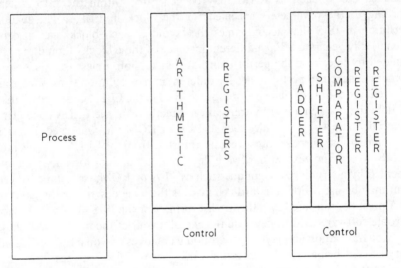

Fig. 3.4 *Module Hierarchy.*

in the layout medium. The ISIS HED editor enables the chip layout to be built in a top-down fashion, the chip outline being filled with 'boxes' which represent the functional parts of the system. Each part can then be filled with further boxes in like manner as the detail is added. Figure 3.4 shows the expansion of detail for some arithmetic process. The highest level box is the process itself, which could contain a register file and arithmetic unit and control element. The register file may have a number of individual registers within it and some control mechanism. The arithmetic unit could comprise an adder, shifter and comparator element. Each would have lower-level components which it is not possible to display here in limited space. It is not essential that the contents of each box is known completely as it is added; a bottom-up editing process can correct errors or optimise the sizes of the allocated parts. By keeping the different abstractions together in terms of their development it is possible to avoid the creation of difficult or infeasible structures in either case. It may be that a library of macro-functions already exists in the ISIS system. In this case some of the regular structures can be created and entered into the floorplan from the outset. This would probably be the case for such entities as RAM and register files. Other items such as datapaths, control processors, ROMs and random logic blocks may be difficult to specify in terms of size until the exact content has been defined through the ELLA route. Estimates of the module sizes have to be entered in these cases, based on previous experience or some more informed method.

As the ELLA code is refined and expanded, more detailed knowledge is accumulated on the previously ill-defined blocks. The structure and size of blocks placed on the ISIS layout can then be further refined also. Regular and predefined blocks such as RAM may be generated within the ISIS system according to the parameters specified for the block in ELLA. The textual description in ISIS that accompanies the layout of such blocks can also be automatically generated. Each different instance of these blocks should then be simulated to determine the performance of that version, unless the macro-cell designer has specified performance for different implementation sizes. Where an equivalent set of library elements to those defined in the ELLA code is available, ISIS data may be generated by translation directly from the ELLA code. This is especially true where modules such as PLA or ROM are concerned. In these cases some special purpose software must translate from the ELLA code file into the ISIS layout and textual hardware description. Random logic may be implemented by a structured technique such as PLA or ROM, or a standard cell layout may be used. For standard cell layouts the code describing the gates in ELLA must be translated into the net-list format for the ISIS library and place and route software used to generate the actual layout of the module. All these automatic translation and layout generation techniques rely on a fully validated library of cells being established in the ISIS system which matches those employed in the ELLA model library.

Where a structured module with unique functionality is required (for example, an array processor) the ISIS system is an effective tool with which to construct the detailed mask layout and circuit description. The flexibility of the system enables the structure to be defined, and the detailed layout of the leaf cells assembled and checked against the defined transistor circuit. The circuit can be simulated and performance optimisation and area minimisation achieved. The implementation can then be checked against the ELLA definition to ensure compatibility at that level. In this manner the basic library available to the designer can be extended to allow the higher performance potential of custom design to be achieved. Such cells and modules may have an identical circuit structure to another in the library but with a different connection philisophy, enabling a saving in connection area or length in the particular application. In other cases a novel logic function may be required. The ability to add special modules of this type within a framework of structured design gives the system the capability of producing designs which are high performance as well as of high complexity.

When custom modules are created or a higher-level collection of connected modules is generated, it is possible to verify the equivalence of the network description and the layout to ensure that the simulated data is equivalent to the layout which exists. This net checking can be applied at any level in the hierarchy provided that all lower levels are in existence.

The management of complexity within the ISIS system is to some degree inherent in the hierarchical nature of the partitioning strategy. It may be further aided by the use of a directory structure which reflects the same hierarchy and contains the data explosion associated with the design. Figure 3.5 shows such a directory. Libraries of a particular type, e.g. RAM, are held in the top of a sub-directory which contains the instance calls of that macro-function. Global libraries of cells must be at the highest common level from which they will be called. As with ELLA, the HDL descriptions must include sufficient comment to describe the issue status of the module and perhaps hold information on the results of simulation and other tests. Control on the issue of cells into a central library can be maintained through such a directory and file data standard.

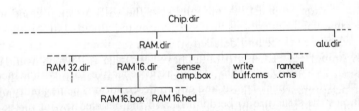

Fig. 3.5 *ISIS Directory structure.*

When a complete and verified system description in the ISIS HDL language and an isomorphic layout has been created, the final checks can be carried out to ensure that the two descriptions are equivalent. Simulations can be carried out to generate the test vectors and the layout data checked for overall compliance with the silicon process design rules. Finally the layout can be processed to produce control data for the mask-making machines, either pattern generator or electron beam. For very large circuits the test vectors may need to be built up from the results of part simulations of the circuit, particularly where built-in test facilities provide special test interfaces. These activities should complete a controlled and verifiable route from system requirement to silicon manufacturing information.

3.5 Developments

Advances in VLSI technology lead to the need for design aids which can handle the greater complexity. In addition new facilities may be required which will enable different and perhaps deeper analysis of the design features. The growth in design complexity has already reached the point where it is extremely difficult for the engineer with expertise in the system aspects of the design to appreciate the constraints of the silicon process. Each extension in complexity leads to the need for some additional constraint or design rationale to be applied to the design process if it is not to become a 'black art'. It is therefore difficult to predict all the extensions to the facilities that could be required for the tools discussed above. Some are relatively easy to see and are already under discussion or development by those responsible for the software production.

There are some processes whose complexity requires abstraction above the structural definitions of a physical realisation, and ELLA should ideally enable the designer to explore these regions with freedom. One medium which may assist in this process would be block schematics, where each block may represent a system state or process, rather in the manner of a software flow chart. Most designers start by describing the process in this manner rather than immediately expressing the algorithm mathematically. Translation from this representation into ELLA suitable for simulation must be provided, perhaps using an expert system approach to add detail.

The translation from ELLA modules into the ISIS medium could also be made less manual. Some of this has been automated already but it is difficult to express the performance and detailed structural requirements in ELLA as it is currently supported. ELLA 'attributes' have been discussed which would enable the designer to select one of a number of ISIS library modules which realise different performance for the same function. Some of the ELLA functional descriptions could also be optimised in terms of logic utilisation. Routines exist which carry out this function and these could usefully be integrated into the ELLA environment. It is not sufficient merely to translate from an ELLA description into a gate net-list or a PLA map if the result would be sub-optimum. These concepts may be to some extent architecture-sensitive, and it is possible that the interfaces into ELLA which already exist need to be used by the architecture developers in order to provide these and other facilities through special software routines.

ISIS has been recently extended to enable parameterised cells such as ROM and RAM to be described. Place and route facilities also exist. As in the case of

ELLA architecture, specific tools are required to enable the efficient construction of elements such as data-paths, state machines and controllers. Another area which has not been addressed is the concept of fault coverage estimation. With the advent of modular design and built-in self-test it will be common for the designer to limit the scope of specific test patterns to relatively small areas. The tests applied to these modules may attempt to be exhaustive. Some measure of the quality of test achieved should be available to the designer; otherwise faults may lie undetected in chips passing the test, with potentially costly effect. No facilities are currently provided toward this end either in ISIS or ELLA.

A restriction is placed on the user of the ISIS system that all blocks and modules must be rectangular in shape. Particularly at the higher levels, this can sometimes be a wasteful constraint. Removal of this restriction would however, also take away some of the conceptual simplicity of the ISIS layouts which is an aid to correct designs. Another restriction which falls into the same category is the difficulty which faces a designer trying to introduce global routing across cells containing specific logic circuits. In ISIS, the strict hierarchical construction rules which contribute to the low error rate require that each cell with a different wiring pattern must be uniquely named and created. This prohibits use use of the ISIS for gate array patterning but also makes the provision of global routing within the overall structure a matter of global planning. With the introduction of three- and four-layer metal processes this could become a major restriction to designer freedom. Some policy would have to be determined concerning the placement and availability of the vias to the underlying circuits but this should not be too difficult to define. At present this is a minor constraint, since it is important to plan power and clock routing within a system for minimum resistance or capacitance according to the needs of the design. A strategy within the bounds of the available technology must be determined which meets the required system performance.

3.6 Conclusions

The author has described a system that he has used since 1984 in the specification and design of VLSI systems and circuits. It has been used for the design and layout of circuits having complexities of from 20 000 to over 300 000 transistors with a very high degree of success at the first process run. As design complexity increases this success rate may be harder to match, but if user experience and the power of the tools grow accordingly it is hoped that the standards can be maintained. The system has shown that it is possible to develop complex and high-performance CMOS VLSI devices which are application specific and for the development costs to be reasonable in terms of the resultant benefits. A key element in the containment of those costs is the ability to produce functional circuits at the first processing attempt. It is a common experience in this field that other parts of the system must be interfaced to the VLSI circuit before the whole system development can be completed. The other elements could also be complex (for example, a microprocessor system), and for the whole programme to be successful and timely it is necessary that working VLSI devices are available as soon as possible. The ELLA interface enables system designers and silicon implementers to communicate through a well-defined and formal interface, which reduces the possibilities of errors of understanding, a common source of difficulty in a complex system specification. The combination of these high-level tools and a silicon design system appropriate for the construction of large and complex custom devices produces a quality result. As further facilities are developed the system should cater for the largest custom designs envisaged in the near future with equal efficiency.

Chapter 4

SILVER: An Integrated Composition System For VLSI Design

D. J. Rees, N. W. Bergmann, J. E. Proctor, N. J. Rothwell

4.1 Introduction

The advent of VLSI circuit design with layout densities in excess of one million transistors per chip is necessitating the research and development of new automated design tools to manage this complexity. Semicustom design methodologies such as gate array and standard cell, together with their appropriate automated design tools, are widely used for integrated circuit design but do not achieve the layout efficiency and performance of full-custom designs. The research being conducted at Edinburgh University is concerned with the development of automated design tools for full-custom design. A structured approach to design is seen as the solution to managing VLSI circuit complexity.

Structured design is not a new concept. It has always been used in engineering design and has been seen as a solution to the mounting problem of managing large software systems ever since computers were capable of running such systems. Electronic hardware designers have recognised that this approach could be adopted and applied to integrated circuit designs and that a hierarchical methodology would reduce the complexity of specifying, constructing and verifying designs of typical LSI and VLSI densities. Principally, this approach advocates the partition of a large scale project into manageable modules with well-defined communication interfaces. By using an iterative process of stepwise refinement, modules are further partitioned until a level is reached at which the designer can make no further refinements and the low level module can then be *logically coded*. These subordinate modules are invoked under the control of the superordinate modules within the hierarchy with control signals and input-output values passed between modules via the communication interfaces.

Applied to VLSI design, a structured approach involves hierarchically partitioning the silicon area into manageable functional blocks. Each block will specify both its logical function and any interfaces, or *ports*, to other blocks within the floorplan. This proceeds to a level, the *leaf cell* level, at which the layout is logically coded. Logical coding involves a detailed description of the functionality of the leaf cell in terms of a technology-dependent mask geometric layout. Leaf cells will also describe port information which determines how they are logically and physically integrated to form higher-level composition cells within the design hierarchy. This process of integration proceeds iteratively in a bottom-up fashion with structural and behavioural information percolating to higher levels of the design until the root node of the hierarchy is processed. At this point the design has been completely integrated. Functional and structural verification should be performed at all composition levels of the hierarchy to ensure overall design correctness.

This structured design approach clearly requires the provision of numerous tools within the design environment, i.e. floorplanners, description tools, verification tools, a high-level composition tool, etc. Furthermore, to circumvent the problem of having to handle the interface specifications manually amongst these tools they should be integrated within an automated design system. The SILVER system aims to provide such a design environment.

4.2 The SILVER Design System

The SILVER design system permits design at two levels:

- *At the leaf cell level.* The primitive leaf cell represents the basic functional building block of the design and is expressed in terms of primitive circuit devices and interconnections. External communications are represented as lists of named ports.
- *At the composition cell level.* The composition cell represents the hierarchical structure of the design and is expressed in terms of composition operators and cell operands which may be leaf cells, composition cells or both depending on the level at which the composition cell is specified within the design hierarchy. By bottom-up design, leaf cells are manipulated as primitive objects by the composition tool to form higher level composition cells. By top-down design, composition cells are defined in terms of cell specifications of function name and communication ports which may be further refined to the leaf cell level.

4.2.1 Floorplanning
Within the SILVER system, a design may be represented hierarchically via a graphical floorplanning tool, or via a special form of declaration within the composition language. The graphical floorplanner is a metaphor of the traditional approach of manually subdividing logical areas on a design space representative of the surface of the silicon chip. The interactive graphics floorplanner is presented to the user as a menu of floorplanning commands such as moving up or down levels within the design hierarchy, assigning port names and layers, adding and deletion of blocks etc. and a silicon area upon which they carve the dimensions of the logic blocks. The editor maintains the hierarchy specified by the user as a binary directed graph whose dynamic structure can be

inspected by the designer during a floorplanning session. Output is a hierarchical, textual representation of the tree which denotes the vertical and horizontal composition of each block within the design space. Attributes associated with each block are: block size, block name and a list of communication ports.

4.2.2 Leaf Cell Design

At the leaf cell level, geometric layout may be customised via an interactive symbolic graphics editor, or via statements within SILVER's purpose-built composition language. The geometric topology is expressed in terms of low-level design primitives such as transistors, wires and contacts, which are modelled in accordance with a user specified set of design rules for the desired target technology, i.e. CMOS, nMOS, etc. A textual specification of leaf cells is described at a later stage when we present the style of the composition language.

The graphics editor incorporates many attractive, state-of-the-art features such as pop-up menu subsystems, real-time zoom facilities with editing at arbitrary scales, real-time panning and multiple viewport management. Additionally, there are optional compaction operations which can be applied to leaf and composition cells in an attempt to optimise the density of the geometric layout. Leaf cells may also be geometrically verified. An on-line DRC program is capable of monitoring the geometry of the layout as it is captured by the graphics editor and instantly highlights any design rule violations. In this fashion a designer is intimately coupled to the layout task at hand and can interactively correct layout violations within a single edit session.

The DRC program is table-driven and hence design-rule-independent, allowing designs to be ported across fabrication units with differing design rule sets. A designer is not locked into the editor, and may, if desired, suspend an edit and resume the layout specification at a later stage. Violations are maintained by the system and any cell descriptions containing them will require edit corrections before a mask-level data format file such as CIF is generated. The editor also performs netlist extraction which can be post-processed to ensure connectivity correctness. SPICE output is also generated as a means of simulating the behaviour of the cell.

4.2.3 Cell composition

In the design of the SILVER language, a key observation is noted that cells which are to be composed together are *typed*. The typing is not only intended to encompass the function of the cell but also its geometrical characteristics. In order for two cells to be composed together correctly their types must *match* in some meaningful way. By performing successive *matched* compositions, a complete chip design will be correctly built up. Composition by abutment is widely regarded as a beneficial discipline in VLSI design. SILVER supports this methodology and also allows overlapped cells. In terms of typed composition, edges must match or be made to match. This style of design forms a background for SILVER. From the point of view of composition, the observable characteristics of a cell relate to its connections to the outside world (its ports) and the circuit features around the periphery of the cell. The type of a cell is regarded a collection of sub-types each relating to one of its ports. Geometrical typing consists primarily of the position, size and layer of ports. Functional typing can, for example, identify a port as a power or ground line, a clock line, or input-output line. A designer could also specify the type of a port in terms of many other characteristics relevant to his design.

Just as the type of an object in a programming language can be *coerced* into a different type appropriate to a given context (for example, integer to real conversion) so the types of cells to be composed here can be coerced into compatible forms if necessary. Coercion can take place in functional and geometrical terms. Functionally, for instance. the drive capability of an output port could be adjusted to match the load of a corresponding input port. In geometric terms, coercion can be implemented in a variety of ways. The cells can be *stretched* to make the ports correspond, an additional *wiring cell* can be inserted between the cells or a more *intelligent* perturbation of the cells can be performed which is hopefully more area-efficient. Successful implementations of all three methods have been made in Edinburgh. The last and most interesting of the three, intelligent perturbation, is described later in the context of the 'Move' primitive. Given that the coercion can be performed automatically by the composition system, the designer is relieved of an arduous task. Correctly composed cells are created easily and quickly. The real significance is that the design will be *correct by construction*. He is thus prevented from allowing errors to persist and mistakenly even to reach the fabrication state. Changes to leaf cells will invoke the coercion system during recomposition and thereby retain overall correctness. The designer may choose to recreate his leaf cells to appropriate dimensions if he wishes to get a better utilisation of silicon area but this is not, in principle, necessary.

When corresponding ports have been composed together, an electrical connection has, in effect, been made in the design. This information can be captured in a typed composition system. Complete knowledge of the whole design can be built up as the hierarchical composition takes place. This is important from the point of view of verification and simulation of the design. The circuit does not need to be extracted at the end of the design process.

4.2.4 Integrity and source control

The SILVER system aims to provide a level of integrity to users who are developing IC designs. The various tools which comprise the design system are available to the user via a consistent language-based interface; interactions with the system are performed by entering statements of the language which are immediately executed. Any alterations made to part of the design might have repercussions in other design fragments. In order for the user interface to present the design in some consistent manner, it would be necessary for parts of the source text to be re-executed, and possibly recompiled. These operations come under the domain of a source control program, which is responsible for maintaining consistency within the design system as a whole. Since the design system is language-based, this implies maintaining consistency at the level of the composition language. A simple prototype source control system to achieve this has been implemented.

4.3 The Silver composition language

High-level programming languages are recognised as powerful notations for expressing design abstraction. Language constructs such as iteration, conditional expressions, recursion and parameterisation allow designers to modularise designs hierarchically and express module composition within the hierarchy in a concise and elegant manner. Results have shown that language-based composition tools are extremely useful in managing circuit complexity.

4.3.2 Language philosophy

Silver features a powerful programming language for composing cells into complete designs. It is based upon an algebra in which *composition expressions* can be written. Such expressions specify how composition cells are to be made up from smaller components in terms of their abutment. The values of the expressions may be assigned to variables and manipulated therein. In the language, cells are primitive data objects. There are primitive expressions for creating leaf cells, and primitive operations for renaming ports and composing cells. Since cells are primitives in the language, they may be passed as arguments to functions, and returned as results.

The SILVER language is block structured, allowing local declarations of variables and procedures. Parameteristion means that functions can be written which generate cells according to values passed as arguments. This allows software tools to be written in the SILVER language itself, rather than as stand-alone programs. The language features the usual control structures, such as conditionals, iteration and recursion. Hence, structures in the design (for example, arrays of sub-components) can be expressed succinctly in a straightforward manner. Variables may be declared and assigned values. Such variables may hold intermediate results as well as fragments of any particular cell.

The SILVER composition language is *functional*; programming is done by means of expressions. Cells are defined by associating a name with an expressions representing a cell. For example, a leaf cell may be declared by associating a name with a leaf cell expression. A parameterisable cell is declared by associating a name with a function which returns a cell as result.

The language is, strictly speaking, semi-applicative; assignment is allowed, as are basic control structures. This makes it easier to generate regular, structured cells without resorting to recursion. The language supports mutually recursive functions for generating more demanding structures. It is strongly-typed. A polymorphic typechecking algorithm ensures that any type errors in a declaration are caught before the declaration is executed. This is a variant of Milner's scheme (Milner 1977) modified to deal with control statements and assignable variables.

4.3.2 Language features

A SILVER circuit description starts with a technology declaration, such as:

technology nMOS

This serves to parameterise the grammar of the language with respect to that technology.

4.3.2.1 Leaf cells: Leaf cells are declared in the language by associating names with *leaf cell expressions*. A leaf cell expression consists of a list of technology dependent *components* bracketted by the keywords **leaf** and **end**. For example:

```
dec Cell1 = leaf . . . components . . . end,
    Cell2 = leaf . . . components . . . end,
    . . .
```

In addition, leaf cell expressions may contain conditional and iterative constructs similar to those found in statement blocks (see below). For example:

```
dec Cell =
    leaf
        for i to 10 do
            if i mod 2 = 1 then
                components
            else begin
                components
            end
    end;
```

Components are divided into four classes: *transistors, contacts, wires* and *ports*. The kinds of transistor, contact and wire layer names permitted depend on the technology specified in the technology declaration. Instantiation statements describe the sizes and positions of the components in the cell and the names of the electrical nets their terminals are connected to. Ports additionally have an identifying name for external reference. Ports on different sides of a cell may have the same name without ambiguity. Furthermore, ports may be declared with the same name on the same side of the cell. In this case, a *bunch* is generated.

4.3.2.2 Composition operators: Leaf cells may be composed into more complex cells by means of composition operators. Prefix composition operators perform rotation and reflection of cells. Infix operators perform abutment of cells. The operators are:

@	anti-clockwise rotation by 90°
@@	anti-clockwise rotation by 180°
@@@	anti-clockwise rotation by 270°
|	reflection about the *y*-axis
/	reflection about the *x*-axis
||	abutment left-to-right
//	abutment top-to-bottom
\\\\	abutment bottom-to-top

Composition operators may be nested to any required depth. For example:

```
dec Cell = leaf . . . end;
dec C1 = @Cell,
    C2 = (Cell ‖ Cell) \\ @(Cell ‖ Cell),
    C3 = Cell // @Cell // @@Cell // @@@(leaf . . . end);
```

Note from the above example that leaf cell expressions may be used anonymously, without being bound to identifiers.

4.3.2.3 Statement blocks: A *statement block* is an expression containing control statements. Execution of the control statements yields a value which is returned as the results of the expression. The first items in a statement block are the *local declarations* of variables. Variables are typed by first usage. They can be initialised with a value (and therefore type) at the point of declaration or later by normal assignment statements. Any variable can be assigned as long as the type of value being assigned matches the type of any previous value. Conditional **if-then-else** statements and **for**-loops are available together with grouping of statements by **begin-end** bracketting. For example:

```
A := B ‖@C;
for i to 10 do A := A‖B;
A := ‖B for i to 10;
```

The last two statements above both have the effect of concatenating ten copies of B together left-to-right and assigning the result to A. The latter is just a shorthand form of the former.

4.3.2.4 Port manipulation: Associated with each cell, whether leaf or composition, are four port lists, one for each side of the cell. Composition has the effect of matching the ports on the sides of the cells being abutted. For instance, the composition

A//B

results in the ports on the bottom of A being matched with the ports on the top of B. The matched ports are removed from the resulting composed cell and the side port lists (the left and right sides in the example) are merged. Port lists to be matched are expected to correspond name by name. Cell type coercion takes care of any mismatch in port positions. If two port lists to be merged contain ports of the same name, a *bunch* is generated. A bunch consists of a number of ports grouped under a common name. During port matching, two bunches must agree in name and size.

A port renaming facility is required in order to allow the matching checks to be peformed correctly when cells with disparate port names are to be composed. It is done using the postfix operator and takes the form:

> *oldname* (*index*) → *newname*

where *oldname* and *newname* are port names and the optional *index* specifies a port or ports within a bunch. For-loops can be used to indicate a range of indexes. For example:

> Cell clock → phi.1;
> Cell bus(i **for** i **by** 2 **to** 10) → even.bus;

The first example renames all occurrences of 'clock'. The second treats bus as a bunch and renames elements [2], [4], [6], [8] and [10].

If the '*newname*' is omitted, the port names are hidden. In effect, the port is removed as a component of the cell type. This is necessary to avoid spurious errors — for example, at the end of a power or ground bus where there is no further extension to the bus line to be made.

4.3.3. The compiler

The compiler is interactive; declarations and expressions entered at the terminal are immediately compiled and executed, resulting in instantaneous feedback to the designer. Cell expressions cause the cell to be constructed and immediately displayed. Alternatively, statements may be presented to the system by means of a text editing interface. In this case, compilation errors are pinpointed within the editor and may be immediately corrected; the system recompiles any definitions altered in this way. Incremental compilation provides fast turnaround in successive revisions of a design, permitting designers to experiment on various design implementations with relative ease.

Error detection results in the offending statement being presented to the user via the text editor; the text may then be corrected, and compilation resumed. Run-time errors – i.e., errors occurring during the evaluation of an expression – are presented to the user in the same manner; the position of the error is pinpointed, and may be corrected immediately, after which the phrase is recompiled and re-executed.

A novel mechanism is used for error reporting and recovery. During the compilation of a SILVER specification file, a text editor is used to maintain the text being compiled. Any error detected during either compilation or execution causes control to be passed to the editor. The editor allows unconstrained alteration of any of the text being compiled. After such an alteration, the compiler is invoked to recompile and re-execute any definitions whose context may have been altered by the editing process. In practice, compilation resumes from a safe point, a *landmark*, immediately before the earliest alteration. The compiler identifies suitable landmarks as it proceeds, for instance at the start of every top-level declaration.

Errors fall into two main categories: *compile-time* errors, such as syntax and type errors, and *run-time* errors, such as port mismatches and the use of unassigned variables. Both kinds of error are indicated to the user in the same way. For example, an undeclared identifier (a compile-time fault) will cause the editor to be invoked at the position in the file where the identifier is used. An unassigned variable (a run-time fault) will cause the editor to be invoked at the position in the file where an attempt is made to access the variable. The editing context (including the current screen display and the values of markers) is preserved throughout the compilation process, allowing related errors quickly to be eliminated.

4.3.3.1 The SILVER abstract machine: Once a SILVER declaration has been parsed and typechecked, it may be compiled and executed. For reasons of modularity and efficiency, the parse tree returned by the parser is not used to evaluate the declaration but is instead compiled into a form of machine code which can then be executed to return the values expected. The code generated is the target code for a specially designed *abstract machine*. A number of language implementations use an abstract machine as their target, including ML (Cardelli 1983) and HYBRID (Rothwell 1986a, b). The SILVER abstract machine is designed with the compilation of the SILVER language in mind, and directly supports a number of the language features as well as manipulation of leaf and composition cells.

The SILVER abstract machine (SAM) is a stack machine. It has two stacks: an *argument stack*, used for arguments and intermediate expression values, and the *control stack*, used for saved displays and return addresses. The run-time state of the machine is distributed on these stacks in such a way that the re-ordering of values within the body of a stack is minimised, thus simplifying the abstract code.

Full details of the compiler and the abstract machine implementation can be found in Rothwell (1986a, b).

4.3.4 Interacting with the SILVER system

In developing the SILVER suite of CAD design tools, it was realised that it would be attractive to integrate individual tools to form a design *system* and to present the user with a clean and consistent system interface. This interface permits designers to gain easy access to all of the tools, manages the necessary software environments required by them and maintains cell libraries for new design layouts.

During a single SILVER session, designers may: experiment with design layouts by creating and altering cells with either the graphics cell editing tools or the SILVER language interpreter; textually compose leaf cells to form larger scale designs, express a hierarchical physical abutment of the cell layout of an entire chip using the graphical floor-planning tool; view the layout by graphically plotting cells; and create leaf and composition cell libraries.

The design tools are used in a hierarchical fashion. First, a designer selects what he is going to do (e.g. create a cell, plot a cell etc.). Next, he may specify the manner in which the operation is to be performed (textually, graphically etc.). Finally, he selects a cell to perform the operation on. This flow is neatly captured by a graphical, menu-based user interface.

At the graphics screen, the cursor object can be freely moved around using a mouse into the position desired for the first-level pop-up menu. The menu hierarchy is presented in the same form as a deck of cards whereby, on selection of a sub-command menu, it obscures the menu on the level above it (apart from its identifying header). This is an attractive system in that the screen is not cluttered with unnecessary menus but leaves the option of returning easily to previous menus. The user can conveniently keep track of the path of his commands to his current situation.

At any stage of the designer's interaction he can plot one of the cells he has created. Normally this will take the form of viewing the artwork on his graphics screen but he can also get hard-copy via a translation to Caltech Intermediate form (CIF) and off-line plot. The integration of cell-viewing on the graphics screen and execution of composition statements in the SILVER language makes for very rapid progress through the design process. Changes can rapidly be made and results seen.

4.3.5 Leaf cell editing

The graphical editor named 'Twiggy' allows a designer to create mask-level cell designs. An on-line physical design rule validation process is incorporated into the editor to ensure that the cells are correct as they are constructed. Errors are immediately displayed to the designer. Only Manhattan geometries are permitted at present, but the data structures used by the editor have been designed to permit an extension to non-Manhattan geometries in the future. In particular, the otherwise attractive corner-stitched data structure is not used. Instead, the other common technique using a regular array of 'pigeon-holes' is used. Each pigeon-hole relates to circuit primitives which are either wholly contained within it or overlap the boundary of it. A list structure for each pigeon-hole contains pointers to common object descriptors. The object descriptors are themselves part of a complex list structure which captures their electrical interconnectivity.

The graphical components of the SILVER tool suite were implemented on in-house-developed hardware which permits a very high-level interface to the display system. A dedicated graphics processor is used to evaluate a tree-like data structure, constructed by the graphics user, and to fill in the frame-store during the process of evaluation. The data structure is somewhat akin to the emerging PHIGS graphics standard (PHIGS 1985).

4.4 The Move primitive

The work on the SILVER suite of tools gave rise to the need to coerce cells in such a way that they could be correctly composed together. The 'Move' primitive was developed with this function in mind, by one of the present authors in particular (Bergmann 1985). It has subsequently proved to be of much wider significance and has applications in a variety of design tools. This section therefore relates also to this wide range of applicability.

Advanced CAD systems for the design of integrated circuits typically consist of a large number of specialised programs to perform individual design capture and design verification tasks. Much attention has been focused on the production of near-optimal solutions to individual design tasks such as sticks compaction, design-rule checking, etc.

A different approach is taken with the Move primitive. It will be seen that this single primitive can be used to produce very simple algorithms for geometric design rule checking, compaction of mask geometry, conversion of sticks circuits to mask geometry and boundary checking during composition of designs. The underlying data structure on which it operates will be described together with an outline of the applicability of the primitive to each of the functions mentioned above.

4.4.1 The data structure

As has been described in previous sections, designs may be specified using either the graphics editor or by interpretation of the SILVER language. Designs are described hierarchically in terms of leaf cells and composition cells. Within leaf cells, circuits are described structurally (i.e., in terms of the interconnectivity of transisitors, contacts, ports and the wires which join them). The data structure used to describe leaf cells captures the components' types (e.g., poly-metal contact), positions and their mask *templates* (i.e., their physical realisation in terms of overlapping rectangles of various mask layers). Information about the electrical net to which each part of a device template is connected is also included.

The method relies on a knowledge of the particular design rules in force. However, these are extracted from a suitable table during the course of operations. This allows the algorithms to be parameterised by means of substituting different values in the table for different rule sets as required. The methods are described below in terms of the dimensionless unit *lambda*. The simplistic view is taken that, given correct templates (according to the design rules), only separations then need to be taken into account to ensure compliance with the rules overall.

Each *item* (transistor, contact, port) is thought of as having *pins* (connection points) on certain layers along certain edges of its template, to which wires may be connected. Wires may be horizontal or vertical, and they have either a user-specified or a default width. A wire is deemed to be of sufficient length to reach between the two items which it connects. This is the key observation which simplifies all the subsequent manipulations. Branches and corners in wires are handled by introducing a special item, called a *null* device, whose template is a square of wire material, to which wires may connect. Figure 4.1 shows a circuit decomposed into items and wires in this way.

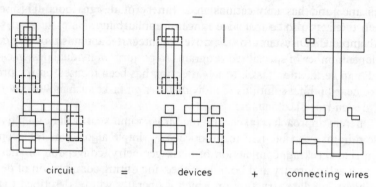

circuit   =   devices   +   connecting wires

Fig. 4.1 *Circuit structure.*

4.4.2 *The operation of 'Move'*

The operation of the Move primitive on the data structure is as follows. To Move a particular item, it is given a particular displacement, in a particular (Manhattan) direction, from its present position. In order to preserve the structure of the circuit, it is likely that other circuit components will also have to be displaced. The operation of the Move primitive is then to calculate the minimum displacement for each other component so that circuit connectivity is preserved, and minimum spacing rules are not violated.

Other components may have to move for one of two reasons. First, if an item is moved, and it is connected to a wire normal to the displacement direction, then that wire will most likely have to move to maintain electrical connectivity. If the wire is connected along an edge of the template which is longer than the wire width, then the item will be able to move up to the limit of this *slack* before the wire has to move with it. Similarly, moving a wire may cause connected items to move. Secondly, if a component is moved, it may come sufficiently close to another component to violate minimum spacing rules, in which case the second component will have to move to restore the minimum spacing. Spacings between items can be calculated by examining the sizes of various layers of the device templates, and the minimum spacings required between these layers. Note that this spacing may depend on whether the two layers are electrically connected or not.

Thus, it can be seen that moving one component may require other components to be displaced by an amount less than or equal to that of the original displacement. This is done by recursively applying the Move primitive, which in turn may require still more components to be moved. In practice, applying the Move primitive does not immediately change the coordinates of the particular component. Rather, a record is kept of the maximum necessary displacement for each component. If a component is required to Move by some amount due to a movement of some other component, then the effect of this move need only be calculated if it is larger than the maximum displacement recorded. Thus, moves are not necessarily recalculated. Furthermore, if a circuit is initially correct, then propagation of Moves will eventually terminate, since at worst each component in a circuit will move by the same amount as the original component.

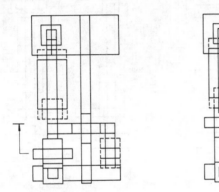

Fig. 4.2   *The Move operation.*

Note that wires in the direction of movement are free to expand or contract to allow the items connected to them to move relative to each other.

The Move primitive is thus similar to cell stretching, since it deforms cells without changing their circuit structure. However, it is argued that Move is more powerful than stretching since it does not necessarily increase the size of cells. For example, in Figure 4.2, the centre port on the left of the cell is moved up by 6 lambda, moving the horizontal wire up by 4 lambda and the pullup by 2 lambda but without changing the size of the cell or the position of the other ports on the same side of the cell.

4.4.3 Pitch matching

Pitch matching is the alignment of connection points in two different cells when those two cells are joined by abutment. The use of the Move primitive in this operation is best illustrated by example. Consider the two cells in Figure 4.3, in particular the centre port on each cell. First, an attempt is made to Move the higher of the two ports down to the level of the other. The results of this Move are calculated, but the positions of the components are not yet changed. Moving a port down may cause other, previously matched, ports to move which is not allowed. Similarly, items are not allowed to move below the bottom edge of the cell. The maximum violation, either moving a matched port or moving below the limit of the cell, is determined and this amount, the *shortfall*, is subtracted from the original Move. This new smaller Move can be made safely. If there was any shortfall, then the lower port is moved up by this amount. This can always be done, since ports are matched from bottom to top, and higher ports are thus free to move. In the example, the centre port on the left cell moves down by 9 lambda and the port in the right cell moves up by 2 lambda.

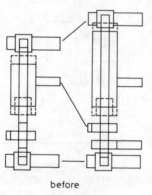 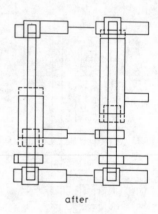

before   after

Fig. 4.3 *Pitch matching of cells.*

Importantly, cells with mismatched ports can be pitch-matched with only a minimum necessary increase in cell sizes. Again, it is worth emphasizing that this is preferable to stretching cells, where ports can only be moved by increasing the size of cells.

4.4.4 Design rule checking
In a system such as SILVER, individual component templates are correct by construction. Geometric design rule checking then reduces to checking the spacing between components. This can be done using Move as follows. Each component in the cell is considered in turn and checks are made in the horizontal and vertical direction. The component is moved by a small distance in one direction, say one lambda horizontally to the East. This Move will propagate to other components, though in this case Moves are not propagated from these secondary Moves. Moves are calculated but the components are not actually displaced. If any component has to move further than the original one lambda to produce correct spacings from the original component, then it must have been too close to begin with, and a design rule error can be flagged. Figure 4.4 shows the result of using the design rule checker on a faulty cell together with the textual description of the error.

The design rule checking algorithm is quite naive — check every pair of components, and see if they are too close — but is very simple to implement using the Move primitive; and, while not particularly efficient, it works at interactive speeds for simple leaf cells.

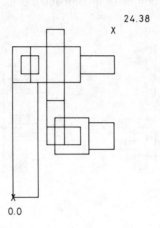

Fig. 4.4   *Design rule checking*
Note: Units and half lambda.
DRC violation between contact at 4,30 and transistor at 10,26 horizontally by 2 units.
DRC violation between transistor at 10,26 and contact at 4,30 vertically by 14 units.
DRC violation between wire at 3,0 and contact at 14,14 horizontally by 2 units.
3 Design Rule Violations.

4.4.5 Compaction

Move can be used to compact designs (i.e., reduce the spacings between components as much as possible without violating design rules). This compaction is one-dimensional, and complete compaction can be done by applying the same process first horizontally and then vertically, or vice versa. Consider the case of vertical compaction. The positions of components are considered above the lower limit of cell geometry, the baseline. Each component is moved downwards just far enough so that the top edge of the component template would lie on the baseline. Such movements are calculated for all components before any changes to positions are actually made. Now, notionally, each component would lie with its upper edge on the baseline. In reality, however, moving some components down to this point will cause others to move even further below the baseline to preserve the minimum spacings between components. Thus it is as if the entire cell had been translated below the baseline, except that spacings between components are just enough to satisfy the design rules (i.e., the cell has been compacted).

It is advantageous to move components down to the baseline starting with the components at the top of the cell, since moving the top components is likely to move most other components a long way below the baseline. When it is the turn of the lower components to be moved, the displacement recorded for them will already be more than that to move them to the baseline and so no further calculation will be necessary.

Figure 4.5 shows an example of a compacted cell.

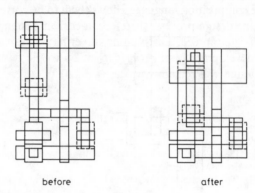

before   after

Fig. 4.5   *Cell compaction.*

4.4.6 Sticks compaction
SILVER allows cells to be described symbolically in terms of virtual coordinates, using a virtual grid, as well as physically. Move can be used very easily to convert such description into mask geometries. Firstly, a template is generated for each component. These templates are placed at positions determined by the virtual coordinates. To ensure sufficient spacing between components, virtual coordinates are all multiplied by some large number, say twice the maximum size of any template plus the maximum spacing necessary between any two layers. This gives a sparse mask-level circuit corresponding to the original symbolic description, which can then be compacted using the algorithm described above to give a compact mask-level circuit.

4.4.7 Boundary checking during composition
Automatic composition systems often require that components be placed sufficiently far away from cell boundaries, so that when cells are abutted, there is no chance of minimum spacing rules being violated by components near the edges of the two cells being abutted. This can be wasteful of space, especially for very small leaf cells which are replicated many times.

By using the Move primitive, components can be placed as close as desired to the cell boundary. When cells are to be abutted, components in one cell can be Moved up to their position next to the other cell, and any components in the other cell which are too close to these components will automatically move away from the edge to preserve correct spacings.

The Move primitive is thus a very neat component of the SILVER 'correct-by-construction' design methodology.

4.5 Conclusions

The SILVER composition system has been shown to be a practical environment for VLSI design. The integration of all the tools, in a systematic way, is particularly useful. The leaf cell editor allows correct leaf cells to be conveniently generated and the composition language allows them to be correctly composed. At each stage of the design process, therefore, a correct design with respect to the design rules, is guaranteed. Last-minute tinkering will not impair the correctness. Design development can proceed right up to the deadline with confidence that the design will remain correct.

References

BERGMANN, N. W. (1985): 'MOVE — A Useful Primitive for a Variety of IC CAD Tools,' *European Solid State Circuits Conference*, pp. 178–85.

CARDELLI, L. (1983): 'The Functional Abstract Machine,' *The ML/LCF/Hope Newsletter*, 1(1), Jan. 1983.

MILNER, A. J. R. G. (1977): 'A Theory of Type Polymorphism in Programming,' *Jo. Comp. Syst. Sci.*, 17, pp. 348–75.

PHIGS (1985): 'Programmers's Hierarchical Interactive Graphics System,' Draft Proposed American National Standard.

ROTHWELL, N. J. (1986a): 'The Generation of Concurrent Code for Declarative Languages.' *PhD Thesis*, University of Newcastle.

ROTHWELL, N. J. (1986b): 'SILVER — The Language and its Implementation,' *University of Edinburgh Computer Science Department Report CSR-206-86.*

Chapter 5

ELLA* — A CAE system for truly hierarchical IC design

C. Miller

5.1 Introduction

The paper describes how the CAD tool ELLA can be used to manage the design of large and complex systems-on-a-chip. The key to solving the complexity problem is a hierarchical top-down design method. Even with a hierarchical approach, a large and complex design comprises a vast amount of design data. A designer's creativity and the quality of the design process can be severely compromised if the designer does not have access to a support environment to aid in the partitioning and management of the design.

5.1.1 The technology explosion

The design of digital electronic systems is becoming progressively more complex as silicon processes improve. At present full-custom chip designers can use several hundred thousand transistors on a chip while semi-custom designers have access to five thousand to ten thousand gates. In the early days of chip design, the main problems were the physical design of the chip, the layout of the chip and the interaction of layout with the lowest level of circuit design. Now the physical side of design is no easier but in addition there are severe problems in getting the system design, chip architecture and the logic design correct. This position was foreseen by the Royal Signals and Radar Establishment (RSRE) in 1978 when they embarked on the development of the ELLA Design System.

5.1.2 Design objectives

The main design objective (Morison *et al.* 1985) of the development team was to produce a tool to aid the design of large and complex systems. The approach taken to control design complexity was to produce a design aid that could be used at all levels of abstraction, from architectural evaluation to low-level circuit design and description. Furthermore, the tool should allow easy mapping between descriptions at different levels of abstraction. Hence the complexity problem could be 'divided and conquered' by a hierarchical approach. Looking

* ELLA is a registered trade mark of the Ministry of Defence

to the future, the designers decided that the system should support the expression and simulation of systems comprising millions of transistors. Finally the inventors did not want the tool itself to restrict an individual designer's style or design methodology, or to constrain the architecture produced. They foresaw that design entry via graphical schematic capture would become totally uneconomic as the size of circuits increased. They also recognised that text can best express abstract behaviour. Text is also precise and often very concise. The ELLA design system was therefore based on a hardware design and description language (HDDL) as the method for design entry.

After more than ten years development ELLA (Praxis 1985) now consists of the ELLA Language, the ELLA simulator, an integrated ELLA support environment and an increasing number of design-support tools.

5.2 The ELLA language

5.2.1 The ELLA hardware model

All programming languages use some sort of hardware model to define the meaning of the language. For example, PASCAL needs the notion of a store to enable use of variables, and a program counter is implicitly required to steer the path of program execution through the control structures used by the programmer. By contrast ELLA uses a lower hardware model which is based on the real world of concurrently operating hardware. In ELLA, a circuit is described as a network of interconnected nodes; each node has one or more inputs and outputs and each node relates to an ELLA function declaration. A function declaration defines those transformations of the input signals which produce the output signals. This hardware model enables the ELLA language to describe both the behaviour and the structure of the circuit being designed. Since a function declaration may itself consist of both structure and behaviour, the problems of mapping between circuit descriptions at different levels of hierarchy are simplified so that, by extension, the hardware can be thought of as a hierarchy of networks, Figure 5.1. The four main features of this hardware model (concurrency, hierarchy of networks, behaviour and structure) combine to give the designer confidence that his most abstract ideas, if expressed in ELLA, can be implemented in silicon.

5.2.2 ELLA signal types

ELLA allows the designer complete freedom to define any type of signal he requires and they can be as abstract as he wants. For example, at the top of an ELLA-text file the following signal types could be declared:

```
TYPE opcode = NEW (load|store|jump|add|boz|nop),
     address = NEW a/(0..65535),
     instruction = (opcode, address),
     bool = NEW (t|f|x|z).
```

The first type, *opcode*, describes all possible operations for a reduced instruction set computer, RISC. Use of these abstracted values enables a designer to delay making a decision on how many bits to use for an opcode until later. Also simulation becomes much more meaningful since the simulator will accept any of the values of opcode as a valid input and, if that input is monitored, 'load', 'store', 'jump', etc. will be printed out together with any other outputs required.

The second type, *address*, has integer values in the range a/0 to a/65535. These integers are called ELLA-integers and cannot be confused with integers which define circuit structure since each integer is prefixed by ' < identifier > /': in this case 'a/'. An ELLA-integer may be used to select dynamically elements in a list, row or structure; either of circuit components, such as busses or memory, or of signals which have been grouped together.

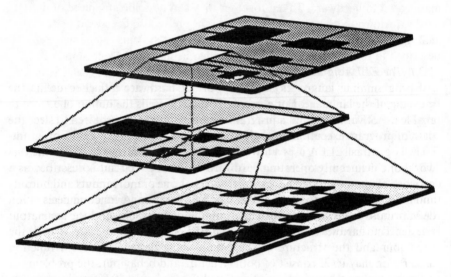

Fig. 5.1 *The ELLA hardware model.*

These first two types are examples of newly defined enumerated types hence the reserved word 'NEW'. The third type, however, is an example of a structure of two previously defined types; hence NEW is not necessary. Using these abstract types, 'instructions' can be processed in a meaningful and understandable way. Moreover, if a particular computer architecture uses the same bus to carry either 'instructions' from a ROM or 'data' to a RAM then it is possible to make the contents of the bus at any one time even clearer by associating, or tagging-on, unique identifiers to either of the two possible types of bus contents. To do this we could continue to define three further types as follows:

```
TYPE int = NEW i/(0..8388607),
     data = (int, address),
     bus_contents = NEW (romad & instruction|ramad & data).
```

Once the design has been decomposed to gate level, then the width of the bus would have been decided and each signal could have values defined by 'TYPE bool' 't' or 'f' or 'x' or 'z'. Even at the bit level the designer has complete freedom. He could have chosen the values as (high|low|dont_care|z): whatever the choice, it is those chosen values which have to be input to the simulator and which would be output from it.

5.2.3 Behavioural primitives

The inventors of ELLA favoured the orthogonal style of language design where a minimal set of general purpose constructs is preferred to a host of specialised ones. This only makes sense if the specialised facilities required can be produced very easily. ELLA only has four behavioural primitives each of which can be mapped on to hardware. Taken together they can describe any piece of digital hardware down to gate-level components. They have also been used to describe analogue circuitry at an abstract level, so that hybrid systems can be successfully described in the ELLA language and verified using just the one simulator.

Only two of the primitives perform value transformations. The first is the CASE clause which acts on any ELLA types, and it is noteworthy because it can be used to describe functionality in a node which is not a leaf node. The second primitive ARITH allows a subset of the functions of a simple ALU to be performed on ELLA-integers.

The last two behavioural primitives are the DELAY and the RAM primitives. The former defines a set of primitives which can be used to store state variables, describe the propogation delays inherent in devices and produce high-level timing functions of the type found in Register-Transfer-Level Languages. The RAM primitive allows the compact description of memory which the simulator is designed to handle both accurately and efficiently.

5.2.4 Parameterised functions

ELLA has a very powerful construct called the macro which takes non-signal parameters as inputs as well as signal inputs. The non-signal parameters can be either integers, (not ELLA-integers), or types.

The use of macros permits the succinct description of generalised hardware, and they have proved very useful in the design of regular two-dimensional structures. Macros can call themselves recursively and this feature is much used for the description of regular non-rectangular structures.

Macro parameters are evaluated statically – i.e., before simulation starts. This means they define the structure which is to be simulated. They can therefore be used with a conventional conditional expression to select which piece of hardware will be simulated.

5.2.5 Algorithmic behaviour

Users of ELLA have the choice of describing hardware using either a functional style of language or a sequential (or imperative) style of language. In the former both the structure of the circuit and the parallel flow of information through it are explicit. Using ELLA sequences, both structure and concurrency are hidden from the user. This may be an advantage when certain types of circuits are described – e.g. register-transfer-level (RTL) architectures and finite state machines.

A unique feature of an algorithmic description which uses ELLA sequences is that it can be transformed to the function-style of language without loss of functionality. This means that the link to hardware has been maintained; a designer can still be confident that his top-level design can be implemented in silicon despite having included sequential code. This very useful feature has been achieved by making sure that the designer can not accidentally build infinite hardware, dynamically reconfigure his circuit or create unplanned side effects – all of which are possible using a conventional programming language for circuit description.

5.2.6 Advanced language features

ELLA has a feature which we call the function type and which is unparalleled by other HDDLs. It allows the designer to describe bi-directional signals as a single ELLA object. Thus, it is used in the description of both a bundle of wires and a single wire in which information flows in both directions. It is an extension to both the ELLA type and function mechanisms, which have been already described.

Some typical uses of function types are to describe busses, tri-state drivers, and transmission-gate logic primitives. Function types are defined by the designer and, as with other types, can range from describing abstract signals to bit-level signals. This feature exemplifies the greater flexibility of ELLA over other HDDLs, which provide only a limited set of wire types. The function set is an ELLA construct which allows a set of functions to be declared together. It is an essential construct for making full use of function types, because it can be used to describe hardware which transforms the signals described by function types. For example, the construct can be used to model bus resolution algorithms. It is a very powerful construct, which has even been used to evaluate a number of transistor modelling algorithms that can resolve different signal strengths.

5.2.7 Hardware synthesis

The guaranteed route to silicon which ELLA provides gives a designer considerable freedom when describing architectures at an abstract level. This is because he or she does not have to have prior knowledge of how the design will be implemented in order to have confidence in the top-level description. This makes ELLA a very suitable medium for designing synthesis tools. A systems designer can use ELLA to produce a description of his system in the knowledge that hard-ware can be synthesised from it.

5.3 Simulating ELLA circuits

The ELLA simulator does not set any arbitrary limits on the type of circuit or signals that can be simulated. It will handle any type and value of signal that has been defined and enumerated by the designer using type declarations. Also the input and output of the simulator are the values of ELLA signals at the level of the designer's own choice of abstraction. Thus, hardware descriptions do not have to be refined to near gate-level before simulation can proceed and they can be monitored by the simulator in a manner which reflects the designer's choice of signals.

The ELLA simulator allows high-level, low-level and mixed-level simulations. At all stages of the design cycle, designers are able to evaluate, optimise and verify the correctness of their work. Errors made early in the design can be corrected before significant effort is made in its decomposition. Mixed-level simulations mean that fewer low-level simulations are required, drastically reducing the total amount of simulation time and cost incurred.

The ELLA simulator is unusual in that it operates on fully compiled object code and is therefore often faster than many traditional gate-level simulators. The time model in the simulator is implemented in ELLA using a very efficient implementation of the normal event-setting algorithm.

The simulator has already been used to verify a design of 600 000 gate-equivalents.

5.4 The ELLA application and support environment (EASE)

EASE provides the designer with a sophisticated support environment in which to integrate large hardware designs described in ELLA prior to simulation. The different units of compiled ELLA text required for a particular simulation are assembled and linked together within a system of contexts.

In addition, EASE provides sophisticated utilities for checking the consistency of the signals throughout a complete design. The system gives the details of any inconsistencies found and does not allow the simulator to be run until the design is consistent. This is a very powerful design aid, which ensures the signals in the designer's circuit are consistently defined at an early phase of the design.

EASE is constructed on a purpose-built, high-integrity, database system (Morison and Peeling, 1985) that manages the designer's data on disc in a transparent way. The designer's data is automatically structured for efficient access and configuration control. The designer using EASE needs to know very little of the host operating system although its features are exploited by EASE.

5.4.1 The compiler

The compiler is written using techniques pioneered at RSRE (Morison *et al.* 1985), to give a single pass syntax and semantics analysis. This fact gives very fast compilation with ten thousand lines of ELLA taking just under five minutes CPU time on a DEC VAX 11/780. Compilation is incremental in the sense that a circuit can be modified by compiling new ELLA-text from new input files without having to recompile the old ELLA-text in the original input file. Also if text being compiled redeclares an element that has already been compiled then the original declaration is silently updated.

Modules of ELLA-text can be prepared using any text editor on the host machine. A module of ELLA-text must meet two criteria if it is to compile successfully. First, as with any programming language, the text must be syntactically correct. Secondly, the declarations in the new module of text must be consistent with all the declarations that have previously been compiled into the context. For example, if the compilation causes an identifier in the context to become inconsistent (e.g., if it uses a function whose interface has now altered) then that identifier is marked as such and simulation can not be done in that context until the inconsistency is resolved, either by modifying and recompiling the offending declaration or by erasing it from the context. This consistency-checking by EASE was a major goal in the design of the ELLA system, and it is a very important feature that finds design errors in the earliest phase of a design.

Since the modular compilation system works at the function level, one has only to recompile at that level and error messages can be natural and informative – e.g.,

> function ADDER uses out of date function GATE

5.4.2 *Contexts*

For any design a user's working space is contained in a library. Once in an ELLA library, contexts may be created into which ELLA-text can be compiled. Each context may be considered as a separate work area in the library where declarations which describe related parts of a circuit design are kept. Contexts are described as either simple contexts or compound contexts; how they differ will become clear later.

Each new library is provided with one empty, simple context called 'ella'. This is the default context. In EASE, the screen prompt is the name of the selected context. On entering a new library, the prompt is

> ella←

When re-entering EASE, the system returns to the context from which the library was last saved, either explicitly or implicitly.

The context system is used to partition design data and to integrate a design by sharing the data between contexts in a controlled manner as described below. The data is usually partitioned to reflect the structure of the hardware being described. For example, a library could be partitioned into the three contexts in Figure 5.2. Here there are two contexts which contain identically named delarations and this illustrates that alternative designs can be developed and manipulated within different contexts without the names conflicting.

5.4.3 Communications between contexts

So that functions can be shared between contexts EASE offers importing and exporting mechanisms. For example in Figure 5.3 'FN ADDER' is exported from context 'behavelevel' and imported into context 'alu'. 'FN ALU' could now be simulated using the description of 'FN ADDER' previously compiled into context 'behavelevel'. All the contexts shown in Figure 5.3 are examples of simple contexts. A simple context which is importing functions or macros is distinguished from a compound context by having a single unnamed imports region.

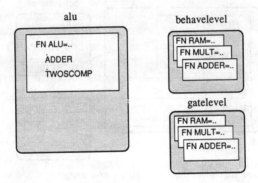

Fig. 5.2 *Design partitioning with contexts.*

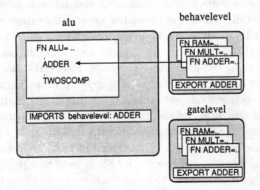

Fig. 5.3 *Sharing declarations between contexts.*

Having proven 'FN ALU' using a behavioural-level version of 'FN ADDER' from context 'behavelevel' the designer could well want to prove the design using a gate-level version. To do this he can convert with a single command the context 'alu' into a compound context with a named imports region – e.g.,

> alu ← nameimports/test1 < RET >
> Context alu is now a compound context with imports region/test1
> alu/test1 ←

Note that the contents of the imports region '/test1' are the same now as they were before, see Figure 5.4. Another command establishes another imports region – e.g.,

> alu/test 1 ← makeimports/test2
> /test2 is now an imports region
> alu/test2 ←

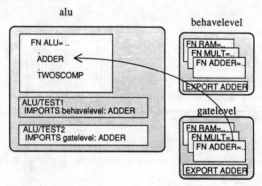

Fig. 5.4 *Simulation of alternative implementations of a design.*

Into this new imports region, the designer can now compile an ELLA text file containing an IMPORTS directive giving the name of the exporting context and the name of the delarations to be imported from it. In Figure 5.4 there is only one import, 'FN ADDER', from context 'gate-level'. Simulation of 'FN ALU' can now proceed using the gate-level version of 'FN ADDER'.

The context system supports designers using top-down (successive refinement) or bottom-up (constructive) or middle-out design methodologies. At the architectural level, hardware elements can be defined behaviourally with an 'empty' body which is to be imported into the design later. Alternatively, a library of hardware primitives can be described in ELLA and can be exported to build high-level design components. EASE is used to handle all the many versions of a design within the EASE database.

The system for importing and exporting ELLA declarations between contexts in EASE provides the user with a sophisticated system for performing mixed-level simulations. Successive mixed-level simulations are easy to perform and are supported by the context system. Alternative descriptions of various components can be imported into the simulation context at different levels of abstraction.

5.4.4 *Design consistency checking in EASE*

As already explained in §5.4.1, the modular compilation system facilitates consistency checking at an early stage in the design. EASE also allows the user to check the consistency of potential changes to the design by use of interactive commands to ask what other declarations depend on a given delaration. For example, if one was working in a context called 'srflipflop' then use of the 'printusage' command could give the following response:

```
srflipflop ← printusage NOR <RET>
FN NOR(4) from [PRAXIS.ELLA]NOR.ELT;1
uses: FN NOT,FN OR, TYPE bool
used by: FN SRFF
```

The library version when the declaration FN NOR was last updated is displayed in brackets at the end of the declaration name (each compilation updates the library version number).

There are a number of other built-in checking mechanisms in EASE. For example, a designer is not allowed to erase a declaration from a context if it is required by another declaration within the same context or by another declaration in another context via export.

A simulation of a function in a context is not allowed until the consistency and availability, through imports and exports, of all declarations needed by that function have been checked. There is an interactive command which checks for the consistency of a context – e.g.,

```
full_adder ← printclosure <RET>
Closure of full_adder contains:
context full_adder
FN HALF_ADDER needs importing
FN OR needs importing
End of closure
```

5.4.5 *ELLANET – An interface between EASE and other CAE systems*

The ELLA system has been designed to protect the existing software investments and design styles that a hardware design team may have developed. An important feature in this respect is ELLANET.

ELLANET is a package that provides an open interface to the EASE database. It enables a designer to write programs that access a hardware design described in ELLA and transfer it to another CAE system. The designer can access the ELLA information at several different levels. One typical output of ELLANET is a netlist description of the hardware described within EASE.

ELLANET allows software to be written in Pascal to access the ELLA data structures within EASE. The package is split into a number of modules that provide pre-defined functions, procedures and data structures which can be included in users' programs.

5.5 Conclusions

ELLA enables the designer to control design complexity by providing a system for true hierarchical design. The mechanisms for this are provided not only by the power and purity of the ELLA language but also by the capability and efficiency of the simulator. Last but not least, ELLA offers a comprehensive set of support tools to aid in the management of the large amount of design data that makes-up a complex design.

Acknowledgements

The author recognises that a useable system depends on the work of many people and wishes to acknowledge the work of the ELLA teams at RSRE and Praxis – in particular: the originators of the ELLA system at RSRE, John Morison, Nic Peeling and Tim Thorp; and Alan Williams and his team at Praxis for their work in evolving an excellent research tool into a viable product. The author also wishes to thank John Saunders at Praxis for his help in preparing this article.

References

MORISON, J. D. and Peeling, N. E. (1985): 'A Database Approach to Design Data Management and Programming Support for ELLA, a High-level HDDL', *IFIP 7th Inter. Symp. on Computer Hardware Description Languages and their applications, CHDL '85*, Tokyo.

MORISON J. D., Peeling N. E. and Thorp T. L. (1985): 'The Design Rationale of ELLA, a Hardware Design and Description Language', *IFIP 7th Inter. Symp. on Computer Hardware Description Languages and their applications, CHDL '85*, Tokyo.

PRAXIS 1985 *The ELLA System Overview* (Praxis Systems plc)

Software tools in teaching ASIC design

Chapter 6

Software tools for microelectronics design teaching

P.L. Jones

6.1 Introduction

The majority of new designs of application specific integrated circuits (ASICs) exploit semi-custom techniques of gate arrays and standard cells to enable engineers who are experienced in the use of SSI/MSI components to make a smooth progression to the benefits of higher levels of integration. Design tools are available from most silicon vendors for the OEM engineer to take a systems concept all the way to a single chip solution. Conversions of existing TTL implementations to achieve matching performance more cost effectively are also straightforward. The marketing of design tools for ASICs has thus become a highly competitive field in respect of both hardware and software, since the value-added earning potential of a powerful workstation CAD system more than justifies the initial outlay and ongoing maintenance.

In parallel with recent developments in software tools, several wafer fabrication plants in the UK and worldwide can now offer access to higher levels of integration and lower unit costs, even at small volume. The integrated circuit manufacturers are looking towards an upturn in the volume of designs rather than long production runs to fill their current excess capacity, and so the retooling required for the economic manufacture and test at low volume has already been set in place. The vital ingredient still missing is the predicted large demand from potential ASIC customers. For example despite all the favourable indications that in applications of IC technology customised components should predominate, the annual output of new ASICs in the UK is no more than a few hundred.

In early 1986 UK government funding initiated a programme of education in custom microelectronics, designed to enable an indigenous electronic equipment manufacturing industry to develop and compete in world markets in the 1990s. Only by education and retraining organised on a national scale could the number of ASIC designers be raised to match a shortfall in UK industry believed at that time to be in the region of several thousand. The approach adopted therefore was to provide computer hardware and software for IC design teaching in all higher educational institutions with degree courses in electronic engineering. Within six months over £8 000 000 was spent on electronics CAD. Arising from this, a support infrastructure was established and the Higher Education ECAD Initiative began. As a result, computer aided design is now an accepted feature of every electronic engineering undergraduate course.

6.2 Aims in IC design teaching

It is essential to maintain a clear view of the concepts to be taught in integrated circuit design. It is all too easy to adopt a mechanistic approach to the application of CAD in electronic circuit and system synthesis, whereas educationally, the teaching of a correct design methodology should be paramount. For example, students must be taught how to handle complexity, to achieve optimum performance from a technology, to economise on chip area, and above all to appreciate the need to design for test. Design teaching in silicon is not just to provide awareness. It must instil the instinctive urge to seek a silicon solution to system integration and to achieve that solution correctly first time.

There are many in the academic world who argue for full-custom design teaching on the grounds that the semicustom approach is intellectually unchallenging and therefore educationally unacceptable. Others will argue that physical layout should not be over emphasised since VLSI systems currently in development incorporate many of the layout automation techniques at present used in semicustom. However, such differences of opinion prove advantageous in practice by providing qualified engineers for the electronics industry with the balance of skills needed to adapt successfully to a continually evolving product environment. The ECAD Initiative software provision thus encompasses all design styles from full to semicustom to ensure that the required diversity of individual skills is developed across a number of educational institutions.

6.3 Facilities for ECAD teaching

Quality tools in an undergraduate course are essential to provide a real experience and understanding of the design process. It is important that the student experience in design should match closely emerging trends in industry. For this, there is no acceptable alternative to acquiring industry-standard CAD hardware and software fully supported and maintained.

In selecting CAD packages, the education sector was faced with the same dilemma which faces any electronics designer in industry today; that is, whether it is better to risk the new and exciting, or to opt for the proven but less sophisticated. In the event, the packages chosen for the ECAD Initiative reflect the need to familiarise students with both advanced and more traditional approaches to electronic system design. Unlike industry, educational establishments cannot balance the cost of a package with an increase in profitability, and so initially it was thought there were severe financial constraints on what could be afforded. Most CAD software suppliers, however, recognised that to capture the educational market it made commercial sense to offer unusually favourable terms. As a result, the combination of packages selected are valued in excess of £30 million, compared with an actual spend of less than £2 million. The modest capital sum made available for the ECAD Initiative thus purchased a wide range of professional software with licences for unlimited use in teaching and research.

The additional provision of ear-marked capital for CAD hardware purchase must, however, be viewed as pump priming on which all institutions should continue to build as funds permit. A valuable source of additional revenue for individual establishments for hardware enhancement can be derived through local industry investing in short courses to retrain its engineers in IC design skills. There is at present some resistance from software vendors to the concept of educational establishments providing training courses in the use of particular CAD tools since the vendors themselves see income and interaction derived from this activity as an essential part of their own support infrastructure. However, in time, there is good reason to believe that software vendors may see higher education as the shop window for their products and so provide encouragement for industry to attend regular up-dating and awareness programmes operating on a national scale through universities and colleges.

6.4 Alternatives for future CAD provision

It is essential that the existing pattern of relationships between the educational estabishments as users and the software houses as vendors be altered for mutual benefit to provide the means to sustain the momentum of the ECAD Initiative procurement. Software vendors must recognise the need to maintain their position in the educational sector.

Because there is no financial gain in teaching and research, it can be argued that there should be only a nominal charge for software licences for educational use of packages which are also applied professionally, for profit in industry. The educational sector should be seen by software suppliers in the marketing context of undergraduate exposure and as the proving ground for new products, with collaboration expected in training courses for industry and the exchange of ideas for enhancements.

There is, however, a good case for levying realistic maintenance charges to cover software support, since this incurs significant revenue costs to the supplier. The level of such charges to members of the ECAD Initiative is of course reduced by operating an arrangement through which direct contacts with the supplier are restricted to three or four lead sites that in turn support a large number of member institutions.

A closer interaction is required between CAD development specialists in academia and industry to ensure the long-term support and future procurement of professional software in design teaching. Much has been stated recently about the need for 'technology transfer' for inventions developed in research environments to be exploited by industry for the good of the economy. However, all too often in CAD, concepts and products developed at public expense by UK researchers are launched commercially at prices which (even at a discount) are far in excess of what individual educational institutions can afford. There appears to be no sign of the attitudes which prevail in the United States at Berkeley and Stanford where internationally recognised tools such as SPICE are in the public domain. In the UK educational establishments are unable to benefit collectively from individual successes in CAD research once that research has been transferred to commercial exploitation.

Thus, it appears that future availability of modern professional CAD in higher education will depend very much on how institutions respond to the tools already provided. If software vendors can identify real commercial gain in terms of subsequent increased industrial demand for those products installed in the academic sector, the future development and expansion of the present scheme will be assured. However, for this to happen, members of the ECAD Initiative should take full advantage of what is being offered at present. If, indeed, the predicted benefits to the supplier can be demonstrated, it may even by possible to negotiate even more favourable terms for software maintenance in the future.

6.5 Conclusions

Universities and colleges have responsibility for initial education of undergraduates in the use of software tools and are expected to collaborate with industry to provide retraining of engineers already in post. The provision in higher education of fully supported industry standards in software and computer hardware for IC design is seen as a prerequisite for future success in competing in world markets for electronic equipment manufacture. However, to maintain state of the art software for future use in teaching and research, a change is needed in the current relationship between CAD developers and the educators.

The continuance of the ECAD provision can only be maintained through a closer liaison with the software suppliers who must therefore be left in no doubt of the commercial benefit of the relationship. One hopes that academia and the software industry will rise to the exciting opportunities presented.

Chapter 7

ISIS in the Educational Environment

B. M. Cook and R. G. Forbes

7.1 Introduction

7.1.1 Background to VLSI design teaching

Over the past few years the use of very large scale integrated circuits (VLSI IC's) has increased dramatically. Many commercial organisations now make extensive use of gate array technology, and an increasing number are turning to full-custom devices to meet their needs. It has become apparent that students studying computer science and electronic engineering need to be educated in the design and use of such technology. They should be familiar with the use of computer-aided design (CAD) equipment supporting the design and construction of VLSI devices.

In response to this educational need, we have introduced courses in VLSI design into the undergraduate timetable. Dr Cook in conjunction with Dr D Gollmann (now at the University of Karlsruhe, West Germany) introduced a final-year option into a computer science degree course at the Royal Holloway College, University of London. Dr Forbes and Dr Cook have introduced VLSI design exercises and a new final year option at the University of Surrey.

The importance of such courses has been nationally recognised and an electronics CAD (ECAD) initiative in British higher education instituted to assist in their provision. The negotiation of national licence agreements for CAD software has enabled courses to be started with only the need to purchase suitable hardware, part of the cost for which was also provided by the initiative.

The CAD system in use for VLSI design both at the Royal Holloway College and at Surrey is ISIS from Racal Redac — a powerful system originally conceived and used by INMOS for the design of the transputer. This provides a fully integrated route from design entry to fabrication data. In this chapter we describe some of our experiences in the use of the ISIS CAD system to support IC design teaching.

7.1.2 Purpose of the courses

Integrated circuits are complex devices whose complete understanding requires a spread of knowledge from semiconductor physics to system design. Such a spread of knowledge is to be found in a team rather than in individuals. Nonetheless, a specialist in some aspect of the design process needs to communicate with colleagues in other specialisations, and must, therefore, have sufficient knowledge of other aspects of the design process to enable him or her to do so.

The aim of the VLSI design course is not to attempt to teach a total understanding of the entire design process. We accept that our graduates will form part of a team and aim to train useful team members. We are attempting to do this by building on the students' present expertise and by introducing other aspects of the design process to them.

Modern techniques in VLSI design mirror the methodologies used in structured programming in order to keep complexity and the scope for error to a minimum. The use of hierarchies, modules, libraries and instancing in VLSI design is rapidly becoming the accepted technique for handling highly complex circuits in a methodical manner. Much value is seen in the courses' ability to reinforce these software concepts in the hardware implementations, and in fact one of the reasons for the choice of the ISIS CAD tools is that they make extensive use of such concepts.

Also, by using a full-custom design it is possible to demonstrate elements of other design styles; in particular parts of the design will illustrate standard-cell methodology. Finally, lessons in design-for-testability are better learned if testing is attempted; this requires fabrication of the completed design. To keep the cost of the fabrication to a mimimum a multi-project chip has been designed, and this is described in this chapter.

7.1.3. The CAD equipment

The course timing places constraints on the type of CAD equipment chosen. Not only should the CAD tools be quick in operation; they should also be such that the need to repeat design work as a result of errors not detected at the layout stage is avoided. In addition, it should be possible for a student with a limited knowledge of device physics to make effective progress on the layout.

Given a carte blanche from which to start, a comparison of the available CAD systems for VLSI design led to the choosing of the ISIS system. Those features particularly instrumental in coming to a decision were:

1. Design rule checking is an on-line interactive process which informs the designer immediately, as each artefact is laid, of any contraventions. Furthermore, in the event of contraventions, both the nature of the error and the artefacts against which failure occurs are displayed on the screen. The result of this facility is that layout is a single-pass process which leads to designs which are always DRC-correct in the shortest possible time, without requiring a detailed knowledge of device physics or fabrication details.
2. The hardware description language used by ISIS contains all the information necessary to permit accurate simulation. Specifically, the electrical net and

the device models are fully defined for both switch-level ('logic') and analogue ('circuit') simulation. All device capacitances are calculated by the simulator, and the user has the freedom to enter estimated maximum values for the interconnections. Thus, a complete simulation can be performed in advance of layout, thereby eliminating the lengthy feedback loop, inherent in other systems, of driving the simulator from the layout. This feature of ISIS leads to another benefit particularly significant in educational environments: that all the HDL editing and simulation is performed on relativley low-cost alphanumeric terminals (VT100 or equivalent) without any need for access to the graphics workstation.

3. The sticks-based artefact-driven layout facilities on ISIS result in a layout which contains all of the electrical connectivity of the HDL used during simulation. ISIS also contains net-checking (so-called 'electrical rule checking', ERC) which permits the layout to be checked against the HDL on each leaf-cell or module, thus eliminating inconsistencies between the HDL (the simulator model) and the layout. Furthermore, the HDL contains all the attributes of the active devices (width, length, area, periphery, type, etc.), all of which are compared with the physical layout during net-checking. Thus not only does the layout correspond logically with the simulator model; it also corresponds at the device model level and thereby ensures that the manufactured product meets its performance specifications.

In addition to the above features ISIS supports a design methodology which embodies structured hierarchies and which maintains an isomorphic relationship between the design concept (HDL or schematic) and the physical design (floor plan and layout). ISIS forces the hierarchies of the two to be identical and net-checking (ERC) is performed at each level to forge an inviolable link between the two.

7.2 The computer scientists course

7.2.1 Course timetable
One quarter of the final year was devoted to a 'computer scientists' option, and this period included the time required for fabrication of the designs. This meant that the project work had to be completed in advance of lectures covering many of the theoretical aspects of IC design. The fabricated devices were available for testing during the third term.

The first week of the course consisted of an introduction to the subject, in which the design process was explained and illustrated. The second week's lectures covered the Hardware Description Language (HDL) and the use of the simulator to verify the HDL and examine the performance of small designs. Students were given assignments to write HDL and simulate simple library cells such as NAND and NOR gates. During the third week the layout and design rule checking (DRC) facilities were demonstrated and explained; students were asked to lay out the cells they had simulated the previous week. (The first students learned to use the CAD facilities and created their first design within one hour, including the design rule and connectivity checks.)

Project work began in week three of the ten-week term. Ten students (working in five groups of two) were given projects which typically required 400 active devices to implement (about 100 gate equivalents). Students were required to partition their designs hierarchically, write HDL descriptions, and to simulate them at the switch and circuit levels. Having satisfied themselves that their HDL descriptions were an accurate and complete representation of the functions they wished to implement in hardware, they were allowed to proceed with layout. Throughout the project discussion sessions were held in place of lectures, allowing much mutual assistance between groups and allowing progress to be closely monitored.

Facilities for beginning designs with schematics entry (which is then converted to HDL by the ISIS software) are available. However, computer science students have no difficulty in comprehending structured languages, so HDL was chosen as the design entry method. This meant that the graphics workstation was not needed during the initial design phase; all of the design capture and simulation was carried out on conventional alphanumeric terminals. Furthermore, there was no intensive demand for plotting facilities to record design data since the system printer could be used for all HDL and simulator output.

Most students finished their design work by the end of the tenth week. Pad drivers, alignment marks and identification details were added to the students' designs by the teaching staff during the Christmas vacation. Design data in GDS2 format were required by the silicon broker by mid-January.

Theoretical aspects of VLSI design were covered in the second term using the textbook by Weste and Eshraghian (1985).

Because the third term was very short as far as teaching is concerned it was used to test the fabricated and packaged designs.

7.2.2 Project details and results

Five projects were undertaken, as follows:

1. traffic light sequencer
2. numerical multiplier
3. finite-field multiplier
4. combination lock
5. cascadable BCD counter/7-segment display decoder.

Of these projects three (1, 3 and 5) were complete on time and assembled into chips. The other two (2 and 4) were incomplete but had some finished modules which were taken and placed on to a chip for testing. Unfortunately, some errors occurred in the process of assembling the chips. These were all detectable by ISIS but were not found prior to fabrication because the test software was not used! Only two of the designs (1 and 5) were correctly assembled – both of which were found to function correctly.

Had the test software provided with ISIS been used, the errors in assembly of the remaining chips would have been found and could have been corrected. To avoid embarrassment we should have checked each design properly before submission for fabrication (the errors were due to the staff, not the students!).

The two correctly assembled designs each yielded six working chips whose performance could be tested. The drive capability and operating speeds corresponded closely with the predictions made by the simulator on ISIS.

7.3 The electronic engineers course

7.3.1 First- and second-year design exercises

Although ISIS is an important tool for final-year design projects, this usage has a limitation: only the relatively small number of students who wish to do (or can be accepted for) IC design projects get experience of the system. Students are doing final-year design projects at Surrey, but we have also taken the decision that *all* of our first and second-year students (approximately 120 in each year) should gain some hands-on experience of the system.

We concentrate here on our introductory design exercises. As stressed earlier, the thinking behind the above decision is that all modern-day engineers should have some appreciation of the IC design progress, with its emphasis on simulation rather than physical prototyping. We also hope that carrying out a full-custom design exercise will demystify those 7400-series little black plastic boxes, and contribute to the making of a link in students' minds between semiconductor physics and electronic circuit behaviour. At a more practical level, students gain some experience of a graphics editor, and of the HDL approach to circuit description.

First-year students, as part of their 'design-and-development' course, get two special lectures on IC (CMOS) fabrication and the IC design process. This is followed up by a short introdutory session with ISIS, lasting about an hour. The hardware is described and the graphics editor demonstrated, and students then work in pairs to lay out and netcheck an inverter.

The second-year design exercise is treated more seriously. The exercise occupies a six-hour laboratory 'day' and students are required to do a couple of hours' work in their own time beforehand. The objective is to design, simulate, lay out and validate a simple hierarchical object such as a half-adder. Students are provided beforehand with notes on (and examples of) the use of ISIS HDL and the ISIS Simulator. As part of their preparation, they design an exclusive-OR gate or a half-adder using NAND-gates and inverters, and write HDL descriptions of the NAND-gate (in terms of transistors) and of their chosen higher-level object.

The day is split into two three-hour sessions: one at the graphics workstation, and one at an auxiliary terminal. Students work in pairs, so that two graphics workstations can accommodate a group of up to eight students. By processing two groups a week, we get through the class within a term.

The terminal session is used to enter and syntax-check the HDL description, to prepare and run Simulator command files that will test the logical function of the designed modules, and to explore different operating modes of the simulator. Most student designs of exclusive-OR gates or half-adders produce a glitch on an output when the inputs are changed simultaneously. The simulator shows this up, and the discovery, exploration and explanation of the glitch is a valuable by-product of the exercise.

The workstation session involves: some demonstrations; an hour learning the CMOS-artefacts editor by going through a work-sheet; an hour to lay out a NAND-gate and perform design validity checks; and an hour to use the 'place-and-route' facilities to build up the higher-level object, from the NAND-gate already built and a pre-provided inverter, and to do hierarchical design validity checks.

Thus, in the course of the exercise each student has gone through the whole IC design process, albeit for a relatively simple circuit, and has a design that would in principle fabricate correctly.

Obviously, the electronic component of the above design is trivial. Also, in real life one might build the half-adder directly out of transistors, in a manner that cannot be directly decomposed into gates. However, our choice of a small hierarchical design is deliberate: the students already understand the digital logic involved; the design can be done quickly; and such a design illustrates the hierarchical features of the system. This said, we also keep in a design file a half-adder built directly out of transistors, and use this to show how much silicon can be saved by adopting a 'non-TTL' design style.

Student reaction to the exercise is generally good, confirming our opinion that all students should be exposed to IC design methodology not later than their second year. In particular, the need to decribe circuits in a hardware description language causes no difficulty for electronic engineering students.

7.3.2 The final-year option
The final year of the Surrey BEng courses is option-based, with each student being required to take six 40-hour options. Probably about 40 students will normally want to take a 'VLSI' option. With a class of this size, and our present option structure and CAD resources, hardware and timetabling limitations make it impracticable to include a design-and-fabricate exercise. We have thus introduced a 40-hour lecture option in 'ASIC Design and VLSI'.

This option runs alongside an existing option, which has its main emphasis on fabrication methodology, process technology, the structure of the various types of transitor, and circuit techniques.

Our 'new' option concentrates more on the design process and on the systems aspects of designing on silicon. Topic headings are as follows:

Basic Principles of ASIC Design (20 hours)

- Introduction – background, design routes, ASIC methodologies, the design process
- Basic principles of NMOS and CMOS digital design on silicon
- Aspects of NMOS and CMOS subsystem design
- Testing and design for testability
- Managment of the ASIC design process

VLSI Systems Design (20 hours)

- I/O Structures
- Some larger building blocks
- Memory design
- Case-study – the Mead and Conway OM2 data-path chip
- PLAs and FSMs and their implementation on silicon.
- Clock skew, synchronisation, and self-timed systems
- VLSI processor arrays
- Hierarchically organised machines
- Future trends and their implications
- Design studies – systems design and floor planning for large projects.

Material of this type has been presented in MSc courses at Surrey and elsewhere for some years now – in our case both before and after we got ISIS. Not surprisingly we have found that giving MSc students experience of using modern design software makes the 'VLSI and CAD' lectures much more meaningful.

To some extent, in our final-year option, we can refer back to the student's experience of the second-year design exercise, and we can also provide supplementry demonstrations using both ISIS and other CAD software. However, it remains our ideal that students should carry out a design-and-fabricate project as part of a 'VLSI' option. But, if this were to be in addition to their substantive final-year project, then we feel that we need at least a quarter of the total time allocated to options, to cover both theory and exercise.

Discussion on such issues is currently in progress. As readers may recognise, there is a fundamental underlying issue here, namely: what is the role of the university or polytechnic in the formation of an engineer? In some eyes, the topic of ASIC design lies uncomfortably in that border zone between basic knowledge (which higher education has a duty to provide) and training in skills (which should be the province of industry). But if much of industry is so unappreciative of the merits of ASICs that the UK Department of Trade and Industry finds it necessary in 1988 to run a Custom Silicon Awareness Campaign, is there not a role for the higher education sector to turn out young engineers with an awareness of new design methodologies, and with enough experience, confidence, and determination to encourage their engineering use? We firmly believe so.

7.4 The multi-project chip

Another potential difficulty with the introduction of design-and-fabricate projects is that of cost. Fabrication costs for projects should ideally fall within a normal project budget. A reasonably complex design should be possible and returned chips should be packaged for ease of testing (normal logic breadboards can be used). A scheme for sharing the cost of a chip amongst several projects is attractive. The idea of a multi-project chip was pioneered by Mead and Conway (1980) but their method requires access to bonding equipment which is not always available within a small department. An alternative approach enabling the use of standard bonding arrangements has been sucessfully used by the authors and is described here.

7.4.1 Project characteristics
Within the context of the computer science degree the scale of project found to be feasible within students project time allocation contains between 400 and 1000 transistors and occupies between 1 and 2 mm² on a 2 or 3 μm process.

In order to include several test points within the circuit in addition to normal inputs and outputs it has been found necessary to allow for 10 to 20 I/O connexions to provide inputs for clocks and data and outputs for data and test points.

7.4.2 The chip layout
When allocating a complete (albeit small) chip to each project it is observed that there is a large number of inputs and outputs for a relatively small circuit. The I/O pad drivers are seen to occupy a large proportion of the silicon area; this area is not part of the students design but considerably increases the cost of fabrication.

By placing more than one project on each chip and multiplexing access to the I/O drivers the 'cost' of I/O is shared between several projects and is less of an overhead for each. By careful design the projects may be isolated from each other and from the common circuitry so that even a major design error (e.g. a Vdd to Gnd short-circuit) will not affect the operation of other projects.

The most economic number of designs on each chip will depend on the fabrication route taken; we used MCE's 3 μm route and found a cost minimum at six projects per chip.

The overall effect is similar to a standard backplane with individual circuit boards (projects) plugged in.

The overall size of the chip is 4.0 × 4.3 mm², and it is housed in a 48-pin package. The pre-defined input and output cells are placed on opposite sides of the chip and joined by a central 'bus' channel (like a letter H). The spaces left are occupied by projects, each of which has access not only to the I/O provided but also to two uncommitted pads so that analogue or other special interface circuits may be included in the projects.

Each project has an area of $1030 \times 1350 \,\mu m^2$ in which the layout may be placed up to the edge of the area; normal input and output occupy one short side. For larger projects (funds permitting) two or three of the project areas may be combined by deleting the intermediary guard bands giving respective areas of $2100 \times 1350 \,\mu m^2$ and $3170 \times 1350 \,\mu m^2$.

The I/O drivers provide:

Four inputs – simple inputs for data including versions for

- TTL or CMOS switching thresholds
- Simple or Schmitt-trigger characteristics

Four special inputs – each having two pads

- one with CMOS threshold Schmitt-trigger characteristic
- one with a pullup resistor and TTL threshold Schmitt-trigger characteristics

These may be used as

- a CMOS switching threshold input
- a TTL switching threshold input with pullup resistor
- a reset input (using an external resistor and capacitor)
- a clock generator in either of the modes
  - a simple oscillator (with an external resistor and capacitor)
  - a crystal oscillator (with an external resistor and quartz crystal)

Twelve tri-state inputs/outputs

- The state and direction can be controlled by the project in three nibbles each of 4 bits.
- The outputs will drive CMOS, TTL (two standard loads) or LEDs (to 20mA).
- The inputs have versions for:
  - TTL or CMOS switching thresholds
  - Simple or Schmitt-trigger characteristics

7.4.3 Test run

The structure and cells described have been designed, combined with experimental projects and fabricated via MCE on a $3 \,\mu m$ single 'poly, double metal process.

The chips worked, first time through, as predicted by the ISIS simulator with regard to switching thresholds, drive capability and speed. The maximum clock speed through the harness is in excess of 30MHz, this being adequate for undergraduate projects.

Yield for the project harness was about 80%. The overall yield for any project will be less than this, but a defect in one project will not affect another with the result that most chips have some working projects; several working examples of each project were found on the test chip.

By using this multi-project chip with six projects per chip and two students per project the cost per student is reduced to something close to a normal final-year project budget, giving us the opportunity to offer a design exercise including fabrication and test to undergraduate students.

The chip was designed with ISIS in eight weeks. Micro-Circuit Engineering (MCE) of Tewkesbury (UK) fabricated the test chip from GDS2 data.

For a standard fabrication run we would obtain 20 packaged chips in about 10 weeks and expect several working examples of each project.

The multi-project chip has been used for undergraduate projects by the University of Karlsruhe under the direction of Dieter Gollmann in the faculty of Informatics.

7.5 Conclusions

ISIS has proved to be an excellent tool for teaching VLSI design, enabling several different types of exercise and project to be undertaken. At its simplest a very short introduction can be given and yet allow hands-on use of the system. At its most complex, given a suitable timetable, complete projects can be taken from design through simulation and layout to fabrication and test within a single academic year. The only essential requirement of the students is that they are able to design a system from the components available; they do not need a detailed understanding of the physics of transistor operation or of the fabrication process.

References

MEAD, C. and CONWAY, L. (1980): *Introduction to VLSI Systems* (Addison Westley, New York).
WESTE, N. and ESHRAGHIAN, K. (1985): *Principles of CMOS VLSI Design* (Addison Wesley, New York).

PART 2

Test and fault-tolerance in VLSI design

Design-for-test (DFT) techniques

Chapter 8

An introduction to testing and design for test techniques

P. D. Noakes

8.1 Introduction

The significant progress made in semiconductor processing technology during recent years now enables devices to be produced with in excess of 500 000 transistors. This trend for larger and more complex integrated circuits shows little sign of diminishing. However, there are increasing technological difficulties in routinely producing the smaller active devices required in order to obtain the required packing density and operating speed. The increasing complexity of integrated circuits also poses problems for the circuit designers and test engineers who are jointly responsible for ensuring that the resulting fabricated circuit meets the desired functional and electrical specification. The cost of testing each packaged circuit, once a design is in production, is a significant proportion of the overall cost of production; typical testing costs are of the order of 10% to 20% of the total cost although in some cases they may approach as high as 50%. In addition it is frequently estimated that the cost of finding a fault in a circuit increases by an order of magnitude for a move from one level of complexity to the next; that is, from the integrated circuit level to the printed circuit board level to the system level (Williams, 1983). It is therefore essential that all integrated circuits should be functionally and electrically checked in full before their inclusion in a system. As the complexity of the circuits increases it is important to ensure that the confidence in the quality of testing is maintained whilst the cost of testing remains an acceptable proportion of the total cost. This chapter introduces the problem of testing integrated circuits and suggests design techniques which can be incorporated in a design in order to facilitate cost-effective testing. Although subsequent chapters discuss some aspects in detail the reader is referred to the appropriate references for further details of other topics.

8.2 The testing problem in integrated circuits

From a testing point of view an integrated circuit, once it has been packaged, has to be treated as a black box which has a finite number of inputs and outputs (Figure 8.1). Unlike a printed circuit board, access cannot be obtained to intermediate points in the circuit unless they are brought out to one of the package pins. The operation of testing requires that certain input conditions are applied and the resultant output conditions are compared with those predicted by simulation or those generated by a known good circuit to which the same input conditions are applied. Circuits which are purely digital can be functionally checked by the application of patterns of 0s and 1s (or *low* and *high* voltage levels). Purely analogue circuits however usually require the use of more specialised signal generators or test equipment which may make automatic testing difficult or impossible. Where analogue and digital functions are mixed on a single integrated circuit (and this is becoming more common), it is sensible to provide facilities to enable testing of each section separately.

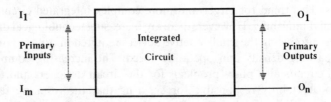

I_1 — Integrated Circuit — O_1

Primary Inputs

Primary Outputs

I_m — O_n

Fig. 8.1 *Representation of an integrated circuit for testing purposes.*

Since the majority of integrated circuits are substantially digital, discussion will be limited in this chapter to testing techniques and design approaches which are applicable to this subset of circuits. In general any digital circuit when flattened to its lowest logic expansion level can be represented as an interconnection of the basic gates AND, NAND, OR, NOR and NOT. Feedback may be introduced round some or all of these gates as illustrated in Figure 8.2. Local feedback is often used to produce bistable elements, whereas global feedback can be used to generate overall time-dependant system functionality.

Testing a digital circuit requires that the circuit is exercised in such a way as to validate its required function and identify any *faults* present in the circuit. Faults within a circuit can take many physical forms and be produced by a variety of problems during processing and operation (Wilkins, 1986). In practice, details of the actual physical cause of the failure is not required in 'go/nogo' production testing and therefore all that need be identified is that a particular *fault effect* is present at the output. For many physical faults, particularly in bipolar circuits, the terminal effect of these faults is adequately modelled by the so called *stuck-at* fault model. A circuit node is said to be stuck-at-0 (S-A-0) or stuck-at-1 (S-A-1) if the application of a pattern at the primary inputs which should drive the node to a logic 1 or a logic 0 level (respectively) fails to produce the expected change at the output. On occasions unwanted shorted and broken (open) connections may produce effects which can not be represented by the stuck-at fault model. It is this type of fault which can lead to fault effects in CMOS structures whereby a combinational circuit becomes sequential as result of the charge storage properties of the gate capacitance (Burgess *et al.* 1988). However, in spite of its limitations, the stuck-at fault model is universally used when discussing the testing of digital circuits.

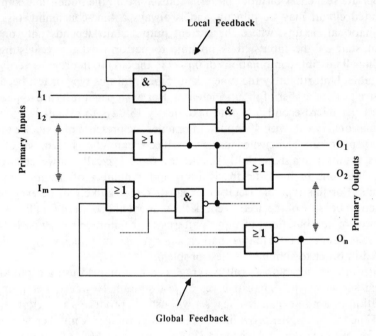

Fig. 8.2 *General representation of a circuit illustrating local and global feedback.*

Functional testing of a digital integrated circuit is concerned with verifying that, when a particular pattern or sequence of patterns is applied to its inputs, it will produce the desired pattern or sequence of patterns at its outputs. Ideally, the applied patterns should exercise (set to 0 and 1) each internal node of the circuit in order to establish whether one or more stuck-at faults exist. The ease of testing a particular circuit, often referred to as its *testability*, is related to how easy it is to control the state of each internal node from the primary inputs (its *controllability*) and the ease of being able to observe the state of each node at the primary outputs (the *observability*). there are two stages to establishing an appropriate set of test stimuli: first, a set of test vectors must be generated; secondly the fault coverage of this set must be evaluated. These aspects are discussed below.

8.3 Test vector generation

When generating sets of test vectors, it is usually assumed that a *single* stuck-at fault is present in the circuit. Assuming that a particular node is stuck-at 0 or 1, a combination of primary inputs is selected which will cause this node to be set to the opposite (1 or 0), fault-free state. Other inputs are set so as to propagate the state of the node being tested to one or more of the primary outputs in order that its condition can be observed. Manual generation of appropriate sequences of input patterns to exercise all the nodes in a complex integrated circuit may be difficult and is always a time-consuming task. In combinational circuits, where the current output state depends only on the current state of the inputs, it is possible to perform exhaustive testing by applying all possible combinations of inputs to the circuit, in order to verify the truth table. Unfortunately the time taken to perform this type of test becomes excessive as the number of inputs increases. For example, if tests are applied at one test per microsecond, it would take nearly 13 days to exhaustively test a combinational circuit with 40 inputs! In practice the presence of a stuck-at fault at every node can be detected using a reduced, carefully chosen, set of test vectors, resulting in a significantly smaller test time. Typically, however, circuits will include one or more feedback loops and a number of bistable storage elements. For this type of circuit an exhaustive test is likely to be even more impractical because of the need to check the sequential operation for all possible sequences of combinations of input patterns to the storage elements. In such circuits, as discussed later in this chapter, a *design-for-test* philosophy should be adopted in order to simplify the test problem.

Sometimes an acceptable fault coverage can be obtained using a randomly selected subset of the exhaustive set of test stimuli. However, deterministic generation of a non-exhaustive test set will usually produce a reduction in the size of the test set whilst providing better fault coverage. A number of techniques, for example the D-Algorithm (Roth, 1966) and PODEM (Goel, 1981, Goel and Rosales, 1981) which are based on the idea of *path sensitisation*, have been developed to allow the deterministic generation of test stimuli. These and other algorithms have been implemented in software and form the basis of current automatic test pattern generators (ATPG). As the task for a large circuit is non-trivial, ATPG programs often require many *hours* of CPU time to produce a set of test vectors. Also, since many of the algorithms used in ATPG programs are not foolproof, automatically generated test vectors should always be evaluated to establish their actual fault coverage. This information can be used to improve the fault coverage by modifying or including additional test vectors. The overall objective is to spend the smallest reasonable time generating a minimal set of test vectors which will give an acceptable fault coverage and which can be applied by automatic test equipment in a reasonable test time. To perform the test generation and application in a reasonable time, the acceptable fault coverage may have to be of the order of 95%, since the CPU time required to generate test vectors to detect the last 5% of faults may be unacceptable. For a circuit of 10 000 gate equivalent complexity, between 15 000 and 25 000 test vectors may be required to give better than 95% fault coverage.

As an example of current trends in ATPG software, Schulz *et al.* (1988) describe the Siemens' ATPG system SOCRATES, which combines several distinct approaches in order to produce what they claim is a highly efficient system for generating test patterns. For example, it incorporates both random pattern generation and deterministic pattern generation; the latter being based on the FAN algorithm (Fujiwara and Shimono, 1983) but including both heuristic and deterministic techniques for improvement. The system also uses a testability analysis to assess the number of test vectors likely to be required in order to give particular levels of fault coverage. These are also used during the decision making in the deterministic pattern generation process.

8.4 Fault simulation

During the generation of a set of test vectors for a circuit it is necessary to evaluate their fault coverage. Although for small circuits this may be performed manually, computer-based techniques must be used for realistically sized circuits. Probably the most common approach currently in use is the fault simulator, which re-simulates the circuit using the desired test set, assuming that there is a single stuck-at fault at a selected node. If the outputs generated during this simulation differ from those produced by the fault-free simulation, the fault is considered to be detected by the particular test set. As the fault simulation must be repeated for S-A-0 and S-A-1 faults at each node, a maximum of $2n + 1$ full circuit simulations are required for n nodes and the CPU time required for a full fault simulation can therefore be considerable. A number of techniques, including Parallel (Seshu, 1965), Deductive (Armstrong, 1972) and Concurrent (Schuler and Cleghorn, 1977) fault simulation and the approximate Critical Path Tracing method (Abramovici, *et al*, 1983), have been used to reduce the required computer CPU time. However, a reduction in CPU time is often achieved at the expense of an increase in the memory necessary for the expanded data structure needed for the circuit description. Even with these improvements the CPU time required to perform a full fault coverage assessment on a typical CAD computer is obviously dependant on the complexity of the circuit, but is often of the order of many hours, if not days!

An alternative approach is to randomly select a subset of the faults and to perform a fault simulation of the circuit using the test stimuli. The percentage of faults detected from the randomly selected subset is then used to estimate the total fault coverage of the test stimuli for all faults in the circuit (Agrawal, 1981). A limitation of this procedure is that it does not produce a list of the faults which are not detected, and the required additions and improvements to the test set may not therefore be clear.

A further alternative to full fault simulation is the statistical fault analysis (STAFAN) technique proposed by Jain and Agrawal (1984). In this technique the probability of detecting a stuck-at fault at every node in the circuit is estimated from calculations of controllability and observability of each node, derived from data produced by a single fault-free simulation. To generate the appropriate data the digitial simulator should ideally be modified to perform a count of the number of times each node beomes a stable-0 or a stable-1 during the application of the test stimuli. In addition, the technique requires a count to be made for each gate input, of the number of times the other inputs are set to a state which allows the propagation of the condition on the particular input. The technique can also be applied as a post-process to the results generated by a simulator provided the stable state of every node can be stored after each test vector. An alternative approach, which does not require the use of sensitisation counters and is therefore faster and requires less memory allocation, has been proposed by Yacoub (1987) and Noakes and Yacoub (1988). In both cases the fault coverage can be obtained and the faults which are likely not to be detected can be predicted. The error as compared to a full fault simulation are of the order of 1% to 4% depending on the circuit and the particular test vectors used (Jain and Singer, 1986, Yacoub, 1987).

8.5 Design for test

As previously stated the testability of a circuit will substantially determine the overall cost of testing an integrated circuit, and it is therefore important for a designer or team of designers to consider how the circuit is to be tested when the overall specification is being formulated (Shaw, 1985). The test strategy must be introduced at the highest level and passed down as the system is partitioned into designable subsystems. Different approaches to testing may often be required for different subsystems (Bennetts, 1984). The object is to increase both the controllability and the observability of nodes at the heart of the circuit. The design approaches, which can be adopted, are often classified as an *ad hoc* technique, a *structured* technique or as *built-in self-testing* (BIST). An introduction to each of these techniques is included below.

8.5.1 Ad hoc techniques

If applied carefully, the testability of complex circuits can be improved substantially using simple additions to the circuitry which involve low cost in terms of the extra silicon area required. The first of these is *partitioning*, in which, by adopting a *divide-and-conquer* policy, large circuits are divided using, for example, multiplexers into smaller circuits which can be tested in isolation as shown in Figure 8.3. In this circuit the inter-block communication is broken by the multiplexers so that, for example, the inputs to sub-circuit 1 can be used as direct inputs to sub-circuit 2 and vice versa. The output multiplexers can be used to allow inter-block signals to be monitored at the primary outputs of the other sub-circuit. This technique improves considerably the controllability and observability of internal nodes in both sub-circuits, and enables more rapid testing of the overall circuit. In this example, the test vectors for each sub-circuit can be generated separately although from a speed point of view it would be sensible to merge the patterns where possible.

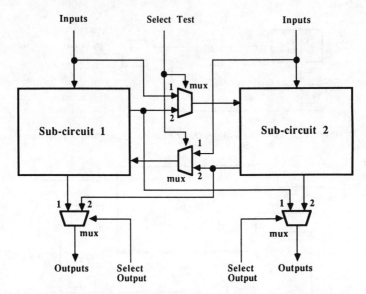

Fig. 8.3 *Use of multiplexers to partition a circuit to improve its testability.*

Whenever possible a fully clocked (synchronous) design approach should be used as free-running (asynchronous) circuits are particularly difficult to comprehensively test. However, where clock generators are included on an integrated circuit they should be de-gated so that an externally controlled clock can be connected to the circuit for test purposes as indicated by the circuit in Figure 8.4. This allows synchronisation of the clock to the test equipment so providing a more controlled test environment. Obviously a facility should also be included to enable the internally generated clock, or a derived related signal, to be monitored. Where possible a master circuit reset should be included in all designs to enable the initial state of all storage elements to be controlled by the test equipment. In addition, it is often useful to provide separate access to allow the state of certain important storge elements to be controlled as required.

When overall feedback is present, a circuit can be more difficult to test; and therefore, it may be necessary to try breaking the feedback paths by using a device such as a multiplexer, as shown in Figure 8.5. Signals from the test generator can then be used as inputs to the broken feedback path during testing.

Finally, when a design is being formulated, it is unlikely for it to require all the pins on the selected package size. It is, therefore, often possible to improve the controllability and observability of internal nodes of the circuit by using the spare pins to provide both additional control and test inputs to other multi-plexers and additional outputs to enable the monitoring of the state of internal nodes. Sometimes the simple insertion of an additional AND or OR gate, or the replacement of a NOT function by a NOR or NAND function, can also be used to advantage to increase the controllability of particular nodes (Motohara, 1984).

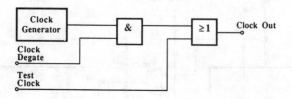

Fig. 8.4 *Isolation of an on-chip clock generator to allow application of an external clock.*

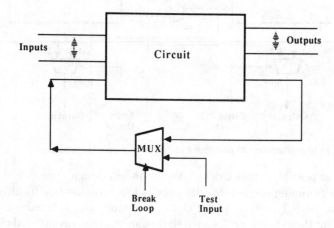

Fig. 8.5 *Using multiplexers to break a feedback loop to facilitate testing.*

8.5.2. Structured techniques
Ideally the techniques discussed above should be planned into the design from the start, but they are often introduced only as an afterthought, when the designer considers how the circuit is to be tested. When using a structured technique, a decision *must* be taken during the formulation of the design to adopt this approach, in order to ensure that the basic architecture of the circuit will facilitate easier testing. In general, a digital system can be partitioned into sub-circuits which, in turn, can often be represented by a block diagram similar to that of Figure 8.6. Recognising that it is easier to automatically generate test vectors for combinational circuits, many of the techniques seek to isolate the combinational logic from the storage elements. Some of the ideas presented

below may be readily applicable to certain parts of a design with only a small overhead in extra silicon area, whilst in other places their introduction may imply a considerable increase in silicon.

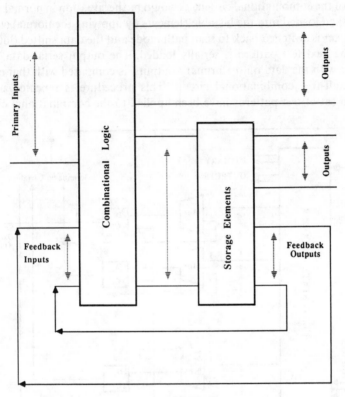

Fig. 8.6 *Finite state block diagram representation of a digital system.*

8.5.2.1 Scan path: One of the most popular techniques introduces a *scan path* by arranging that, for test purposes, the storage elements can be connected into a shift register as illustrated by the structure shown in Figure 8.7 (Williams and Parker, 1983).

The circuit can be tested in two phases. Firstly *scan select* is used to select multiplexer input 2 so as to configure the D-type bistables as a shift register. This is tested by serially loading a sequence of patterns (for example 0,0,1,1,0,0,1,1,...) through *scan data in*, which will cause each of the bistable elements to be cycled through all possible sequences of changes contained in its sequence table. The results of these changes will be shifted through the register so that the resultant serial output pattern can be compared with that which would be expected from a fault free circuit. Any differences indicate that one or more bistables or interconnections are faulty in the scan path.

When it has been established that the storage elements are fault-free, the combinational logic can be tested using a pattern applied to the primary inputs together with a pattern loaded serially into the shift register. The multiplexers are then switched so that the inputs to the D-type bistables are derived from the output of the combinational circuit, as would be the situation in normal use. The data is then loaded into the bistable elements by applying the normal clock, the multiplexer is switched back to scan path mode and the data shifted out serially while the next test pattern is serially loaded. The output serial data pattern, together with the data on the primary outputs, is compared with that expected for a fault-free combinational circuit. This procedure is repeated until the required set of test patterns have been applied to the combinational circuit.

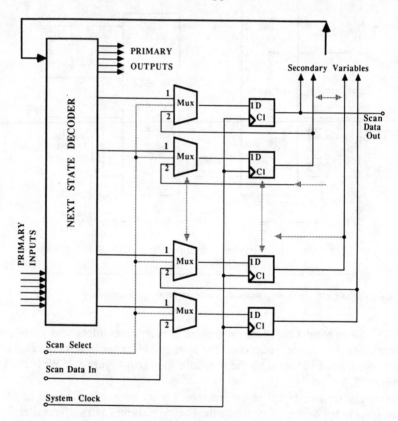

Fig. 8.7 *A circuit configured to include a scan path.*

An implementation of this concept proposed and used by IBM, is referred to as level sensitive scan design (LSSD). This requires the use of the shift register latch (SRL) circuit shown in Figure 8.8. The operation of the SRL circuit has been designed to be independent of the ac characteristics of the clock and requires only that the clock is held high long enough to enable the feedback loop to stabilise. The shift register operation is obtained by connecting the output of L2 to the SD input of the next bistable and then clocking the A and B inputs to produce two-phase clocking (Figure 8.9). Further details of the use of this particular circuit are included in Eichelberger and Williams (1978).

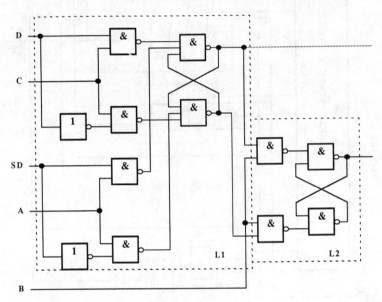

Fig. 8.8   *IBM shift register latch.*

The scan path technique provides a significant improvement in terms of the ease of testing but incurs an overhead in silicon area due to the need to include the extra multiplexers and inter-connections required to produce the scan path structure. In addition up to three extra input/output pins may be required to enable the testing to be performed. Also there is a time penalty introduced because of the need to serially load the new test vector as the results are shifted out.

In an attempt to reduce the overheads introduced when all the bistables are required to be connected in a single long shift register, a number of proposals have been made to adopt a partial scan design strategy. As the name suggests, when using this philosophy only selected bistables are connected in the scan path. There is an obvious reduction in the ease of testing using this technique. The main difficulty is in determining which bistables are to be used in the scan path. Agrawal *et al.* (1988) propose a technique which involves using a modified PODEM test generation program for the combinational logic. Rather than just generating one test per fault, this program generates all the tests for each fault. For each test only those bistables which produce inputs which are essential for detecting a particular fault are set to 0 or 1; other inputs are set to 'don't care'.

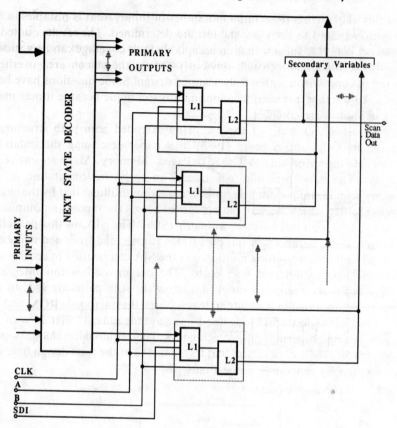

Fig. 8.9 *Use of SRL circuits to produce a scan path.*

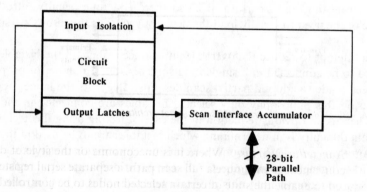

Fig. 8.10 *Principle used in the Motorola MC68851 to enable embedded circuitry to be tested using a scan path.*

From this information the minimum number of inputs, that is bistables in the scan path, required to carry out the test are determined. The results quoted in Agrawal *et al* (1988) suggest that an acceptable fault coverage can be achieved using this technique, which will obviously reduce the silicon area overhead incurred in comparison with a fully scanned design. Some questions have been raised as to whether it is sensible adopting this technique because it may make test generation more complex.

In contrast, an example of the use of an extended scan path structure is discussed by Giles and Scheuer (1986) in a paper describing the testability features of the Motorola MC68851 Paged Memory Management Unit (PMMU). The basic principle, which uses an early implementation of the boundary scan technique discussed in a later section, is illustrated by the circuit in Figure 8.10 in which the scan path is routed across the inputs and outputs of the circuit to a scan interface accumulator (SIA). The SIA, in this particular circuit, is a 28-bit parallel to serial port which can be read from and written to by the ATE. By incorporating multiplexers the SIA can be used to control and interrogate various different scan paths. The design used in this integrated circuit provides an interesting insight into how the scan paths are used to test the PLAs incorporated in the finite state machines, the microcode ROM and the Execution Unit. The statistics quoted by the author indicate that 75% of the chip area is tested by structural test techniques. This required less than 5% extra silicon area and the authors claim that less than 5% of the total design time was spent on actually generating the test patterns.

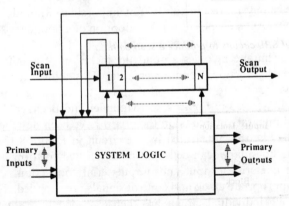

Fig. 8.11 *Block diagram illustrating basic scan-set architecture.*

8.5.2.2 Scan-set architecture: Where it is uneconomic or the style of design makes it inconvenient to introduce a full scan path, a separate serial register can be introduced to enable the state of certain selected nodes to be controlled, and other nodes to be interrogated and their logical state shifted out as a series pattern (Figure 8.11). In order to test the circuit, the register is serially loaded with a test pattern. A number of the register outputs are used as the control inputs to multiplexers which allow inputs to selected nodes to be taken from other outputs of the register. Appropriate levels are also applied to the primary inputs during testing and the logical state of the selected nodes is loaded into the register and the results shifted out as the new test pattern is shifted in. These results together with the states of the primary outputs can be used to assess whether there are faults present in the circuit. The area overhead incurred by the introduction of this testing structure is less than that of a full scan path whilst providing a reasonable improvement in the testability of the circuit. It may also be possible to arrange for this structure to provide the facility for monitoring certain nodes of the circuit during normal use. A version of this structure is provided in the form of the so-called shadow registers in many of the AMD bit slice products.

8.5.2.3 Boundary scan: The introduction of a boundary scan structure within integrated circuits has received considerable attention since the establishment of the Joint Test Action Group (JTAG) in 1985. This group includes representatives from many major European and North American companies who are both manufacturers and users of integrated circuits. The object of the group was to investigate establishing a test architecture and philosophy standard which would simplify the testing of printed circuit boards containing large numbers of increasingly complex integrated circuits, many of which are now surface mounted on both sides of the board to increase the packing density. The use of traditional bed-of-nails test-heads for ATE equipment for such boards has become increasingly more difficult, if not impossible. Maunder and Beenker (1987) discuss JTAG's proposals for the adoption of a boundary scan technique as a framework for structured design for test.

The basic technique revolves around the inclusion of a shift register latch adjacent to each functional component pin. This allows the signals at the component boundaries to be controlled and observed using scan path testing principles, as illustrated in Figure 8.12 (Maunder and Beenker, 1987). Here all the individual integrated circuits on the board have been designed with boundary scan cells like that illustrated by the circuit in Figure 8.13. Within each circuit, all the boundary scan cells can be connected to form a shift register around its perimeter. An input and output serial connection is made to this register and appropriate clock and control signals are supplied. The scan registers within each individual circuit are then cascaded to form a single shift register with a *board serial input* (BSI) and a *board serial output* (BSO). In practice, to enable faster testing, a facility is included which allows the on-chip scan path to be short-circuited, thereby allowing the removal of individual chips from the board scan path.

With this type of structure it is possible to identify three possible modes of testing. First, during an external test mode the boundary cells are selected such that internal circuitry is isolated and external circuitry and interconnections

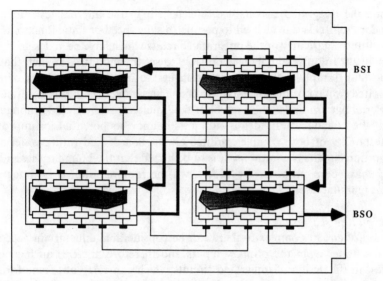

Fig. 8.12 *A board illustrating a boundary scan configuration.*

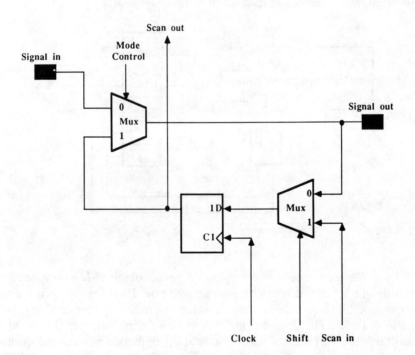

Fig. 8.13 *A suggested circuit for a boundary scan cell (Maunder and Beenker, 1987).*

outside the integrated circuits are fully tested. In the internal test mode the boundary scan cells can be used to perform a slow speed or static functional test of the internal circuitry of the integrated circuits which have been designed with boundary scan cells. Finally it is possible to set the multiplexers such that the latches can sample the states appearing at the inputs and outputs of each integrated circuit during normal operation. The sampled data can be shifted out as a serial pattern under the control of the boundary scan clock. Although the use of the boundary scan idea is still in its infancy, its potential for improving the testability of both complex printed circuit boards and future wafer scale integration circuits is enormous. It is to be hoped that the boundary scan ideas, which have been proposed by JTAG, will be rapidly adopted by integrated circuit manufacturers.

8.5.3 Built-in self-testing

Because of circuit complexity there can be considerable difficulty in accessing internal nodes from the primary inputs and therefore, there is an increasing interest in the design of integrated circuits which can carry out some form of built-in self-testing. For example, Gelsinger (1986) reports that the Intel 80386 microprocessor has been designed with testing in mind and therefore includes a mode which allows testing of the three internal PLAs and the microcode ROM. The techniques used for self-testing this type of circuit are based on a data compression technique referred to as *signature analysis*.

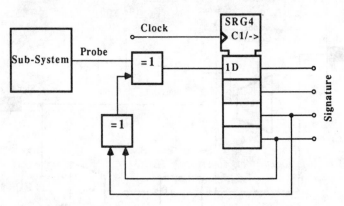

Fig. 8.14 *A four-stage signature analyser.*

8.5.3.1. Signature analysis: Checking the results obtained following the application of a set of test vectors is a non-trivial task. However, by using a data compression technique, the task can be eased by producing a series of so-called *signatures* for the circuit under test. The basic principle involves the use of a linear feedback shift register whose maximal length pseudo-random sequence is altered by the sequence of 1s and 0s appearing at the node being monitored. An example of a four stage signature analyser illustrating the principle is shown in Figure 8.14.

Assuming the same register starting state, a *good* sequence appearing at the monitored node will, for the same number of clock pulses, always cause the register to be left storing the same pattern. This is referred to as the 'signature' for that node. Provided that the nominal pseudo-random sequence is long enough (that is, the register has a reasonable number of stages) the likelihood of the register ending with the correct or *good* signature when a faulty sequence appears at the monitored node, is statistically remote. The principle can be extended to allow a number of nodes to be monitored simultaneously as illustrated in Figure 8.15. In this circuit, the Z-inputs are connected to the nodes being monitored and, by superposition, the final signature will be determined by the pattern of 0's and 1's appearing at the monitored nodes.

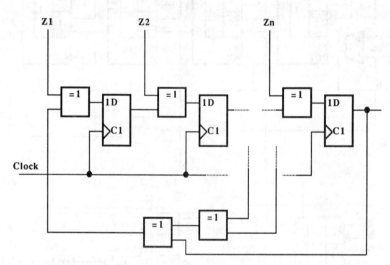

Fig. 8.15 *Multi-input parallel signature analyser.*

8.5.3.2 *Built-in logic block observer (BILBO):* The basic signature analysis idea discussed above has been extended to produce the BILBO (Konemann *et al.*, 1979), as shown in Figure 8.16.

In the BILBO a number of nodes are again simultaneously monitored using the Z inputs, and the sequences appearing are used to modify the sequence of the linear feedback shift register (LFSR) when $B1 = 1$ and $B2 = 0$. Using the principle of superposition, any errors will cause the final signature to be different to that for a good circuit. If a reasonable number of stages are used in the BILBO, there is a low probability of the correct signature being generated when there is a fault present on one or more nodes. If the Z inputs are held at 0, the circuit will act as pseudo-random pattern generator. When $B1 = 0$ and $B2 = 0$ the BILBO acts as a shift register with serial data entering through SD_{in} thus allowing a form of scan path operation. When B1 = 1 and B2 = 1 the data appearing in parallel on the Z inputs is loaded into the register, when a clock pulse is applied, producing a PIPO or PISO register. If B1 = 0 and B2 = 1 the D-types can be cleared by the application of a clock pulse. The BILBO structure can be readily incorporated within an integrated circuit by modifying, for example, an input or output staticising (glitch removal) register. It can then be used to perform a number of functions during self-testing.

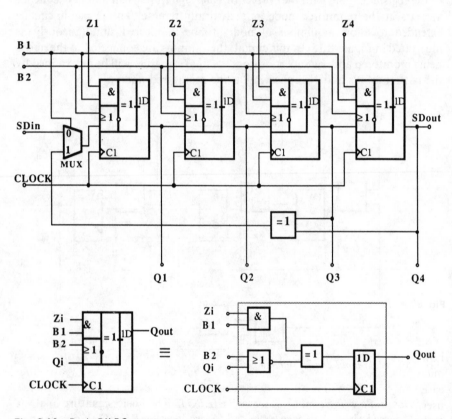

Fig. 8.16 *Basic BILBO structure.*

For example, a separate BILBO or an existing input register may be configured to act as a pseudo-random pattern generator and used to supply test vectors to the rest of the circuit (Figure 8.17). Alternatively, for an embedded register, a deterministic set of test vectors can be loaded in parallel or scanned in using the shift register configuration. A second BILBO or an available output register configured to perform signature analysis is used to monitor the outputs of the circuit to which the outputs of the pseudo-random test pattern generator are applied. In this type of application, the circuit must be simulated to establish whether the random test patterns generated by the input BILBO provide an adequate fault coverage and also to derive the good signature which should be present in the output BILBO at the end of the test.

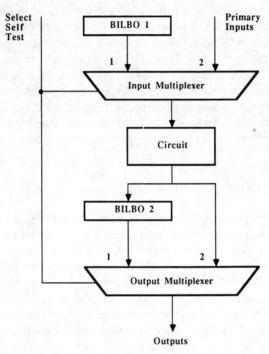

Fig. 8.17   *Example of a built-in self-test configuration.*

8.5.3.3 Self-testing in the 80386 microprocessor: Gelsinger (1986) discusses how some of the above techniques are put into practice in the Intel 80386 microprocessor in order to ensure that the device can be tested with reasonable ease. The approach also allows some aspects of the device to be tested by the user. The architecture of this device is discussed in reasonable detail by El-Ayat and Agarwal (1985). The structure uses three large PLAs to perform various random logic functions and a ROM to store the instruction microcode. Under test conditions the inputs to each PLA are connected to the outputs of an appropriate length linear feedback shift register (LFSR) which is used as a pseudo-random pattern generator to produce a randomly generated exhaustive test set. The outputs of each PLA are connected to an appropriate parallel load signature analyser (similar to Figure 8.15). The microcode ROM is exhaustively tested by cycling the counter attached to its inputs through its complete count sequence while capturing the outputs in a further parallel-load signature analyser. A single-bit output is taken from each of the four different length signature analysers into four separate 16-bit serial input signature analysers.

The outputs of these signature analysers after a 512 000 cycle self-test are added as two 32 bit words in the ALU and compared with a prestored constant. If the result is 0, then it is concluded that the self-test has been successfully completed. The self-test phase allows approximately 52% of the 285 000 transistor sites to be tested. However, Gelsinger indicates that the overhead in extra silicon area is only about 2% of the total chip area. In addition, special functions or instruction codes, referred to as *test hooks*, are included to enable testing of other areas of the device. Finally, specifically generated test vectors are used in order to complete the testing of the 80386 microprocessor. In all the total test set requires about 800 000 clock cycles. To perform a switch level simulation of this device, Gelsinger indicates that approaching 100 CPU days on a IBM 3090 computer are required. To produce a full-fault simulation requires significantly longer and therefore a full fault coverage assessment is currently not possible.

8.6 The cost of including DFT in a design

It is impossible to prescribe a particular design-for-test strategy which is appropriate for all integrated circuit designs, since different architectures lend themselves to the use of particular design for test techniques. However, it must be noted that all of the modifications lead to an increase in the area of silicon required, and a designer must therefore estimate, while formulating the design, the extra silicon area which will be acceptable (Miles *et al.*, 1988). The use of some of the so called *ad hoc* techniques will generally involve a smaller increase in area (for example 5–10%) than *scan path* and *self testing* circuitry (up to 35–40% has been quoted). Another point to note is that the presence of the additional circuitry used to improve the testability will often introduce an extra delay in the normal signal paths and may, as a result, reduce its maximum operating frequency. The size of the package may also have to be increased in order to provide the extra pins required to facilitate the control of the test circuitry. In general these factors may result in a higher manufacturing cost because of the larger chip area of the device and the increased probability of obtaining a faulty device. Integrated circuit designers must, however, trade off these factors against the reduction in the time required both to generate and apply the test patterns and the consequent reduction in overall costs that result.

8.7 Conclusion

Designers are increasingly having to consider the use of application-specific integrated circuits (ASICs) in their designs. Whether these are customised gate arrays or full-custom designs, it is necessary for the circuit designer to consider how the final integrated circuit is to be tested. The sensible use of the simplest ad hoc techniques can improve the testability of a design substantially, while the introduction of a scan path allows combinational and storage elements to be tested separately. The JTAG proposal on the use of *boundary scan* is interesting and is likely to have a significant impact on future integrated circuit and board design and testing strategies. Any well-thought-out attempt to provide some standardisation in this area is to be encouraged. Built-in self-testing is becoming more common, and in some cases a necessity, in the complex integrated circuits now being designed.

There is no doubt that a clear *design-for-test* strategy is an essential requirement of the specification if future VLSI and WSI circuits are to be efficiently and economically designed and tested.

References

ABRAMOVICI, M. MENON, P. R. and MILLER, D. T. (1984): 'Critical path tracing: an alternative to fault simulation', *IEEE Design and Test*, **1** (1), pp. 83–93

AGRAWAL, V. D. (1981): 'Sampling techniques for determining fault coverage in LSI circuits', *J. Digital Systems*, **3**, pp. 182–202

AGRAWAL, V. D., CHENG, K. T., JOHNSON, D. D. and LIN, T. (1988): 'Designing circuits with partial scan', *IEEE Design and Test of Computers*, **5** (2), pp. 8–15

ARMSTRONG, D. B. (1982): 'A deductive method for simulating faults in logic circuits', *IEEE Trans. on Comp.*, **C-21** (5), pp. 464–71

BENNETTS, R. G. (1984): *Design of testable logic circuits*, (Addison-Wesley, New York)

BURGESS, N., DAMPER, R. I., TOTTON, K. A. and SHAW, S. J. (1988): 'Physical faults in MOS circuits and their coverage by different models', *Proc. IEE*, **133E**, (1), pp. 1–9

EICHELBERGER, E. B. and WILLIAMS, T. W. (1978): 'A logic design structure for LSI testability', *J. Des. Automation and Fault Tolerant Computing*, **2**, pp. 165–78

EL-AYAT, K. A. and AGARWAL, R. K. (1985): 'The Intel 80386 — Architecture and Implementation', *IEEE Micro*, **5**, (6), pp. 4–22

FUJIWARA, H. and SHIMONO, T. (1983): 'On acceleration of test generation algorithms', *IEEE Trans. on Comp.*, **C-32**, pp. 1137–44

GELSINGER, P. P. (1986): 'Built in self test of the Intel 80386', *Int. Conf. computer Design, New York*, pp. 169–73

GILES, G. and SCHEUER, K. (1986): 'Testability features of the MC68851', *Int. Test Conference, 12.3*, pp. 408–11

GOEL, P. (1981): 'An implicit enumeration algorithm to generate tests for combinational logic circuits', *IEEE Trans. on Computers*, **C-30**, (3), pp. 215–22

GOEL, P. and ROSALES, B. C. (1981): 'PODEM-X: an automatic test generation system for VLSI logic structures', *Proc. 18th IEEE Design Automation Conference*, pp. 260–8

JAIN, S. K. and AGRAWAL, V. D. (1984): 'STAFAN: An alternative to fault simulation', *Proc. 21st IEEE Design Automation Conference*, pp. 18–23

JAIN, S. K. and SINGER, D. M. (1986): 'Characteristics of statistical fault analysis', *IEEE Int. Conf. on Comp. Design*, pp. 24–30

KONEMANN, B., MUCHA, J. and ZWIEHOFF, G. (1979): 'Built-in logic block observation technique', *Proc. IEEE Test Conf.*, pp. 37–41

MAUNDER, C. and BEENKER, F. (1987): 'BOUNDARY-SCAN: a framework for structured design for test', *Proc. of IEEE Int. Test Conf.*, pp. 714–23

MILES, J. R., AMBLER, A. P. and TOTTON, K. A. (1988): 'Area overhead prediction for testable VLSI', *Proc. of EDA Conf.*, London pp. 129–38

MOTOHARA, A. and FUJIWARA, H. (1984): 'Design for Testability for complete test coverage', *IEEE Design and Test*, pp. 25–32

NOAKES, P. D. and YACOUB, M. N. R. D. (1988): 'Statistical fault analysis: Is it an alternative to conventional fault simulation', *Proc. of EDA Conf.*, London, pp. 507–16

ROTH, J. P. (1966): 'Diagnosis of automata failures: A calculus and method', *IBM J. Res. Develop.*, **10**, pp. 278–91

SCHULER, D. M. and CLEGHORN, R. K. (1977): 'An efficient method of fault simulation for digital circuits modelled from Boolean gates and memories', *Proc. 14th IEEE Design Automtion Conf.*, pp. 230–8

SCHULZ, M. J., TRISCHLER, E. and SARFERT, T. M. (1988): 'SOCRATES: A highly efficient automatic test pattern generation system', *IEEE Trans. CAD of Integrated Circuits and Systems* **17**, (1), pp. 126–37

SESHU, S. (1965): 'On an improved diagnosis program', *IEEE Trans. on Electronic Computers*, **EC-14**, (2), pp. 76–9

SHAW, S. J. (1985): 'Test strategy planning', *IEE Design for Test Colloquium Digest (No 1985/98)*, pp. 3/1–3/5

WILKINS, B. R.(1986): *Testing Digital Circuits — An Introduction*, (Van Nostrand Reinhold, UK)

WILLIAMS, T. W. and PARKER, K. P. (1983): 'Design for Testbility — A Survey', *Proc. IEEE*, **71**, (1), pp. 98–112

YACOUB, M. N. R. D. (1987): 'The statistical assessment of the fault coverage of test stimuli applied to digital integrated circuits', *MSc Thesis, University of Essex, Colchester*

Chapter 9

Testability and diagnosability of a VLSI systolic array

W. R. Moore

9.1 Introduction

The systolic array has emerged as an example of an architecture which is well suited to VLSI implementation of arithmetic and signal processing functions (Urquhart and Wood, 1984, Moore *et al.*, 1987). The approach consists essentially of manipulating the problem so that data are clocked through an array of identical cells; the repetition of the cells makes the systolic array quick to design and verify and the clocking of each cell avoids the timing problem associated with long lines across a chip.

An equally important factor in VLSI designs is the ability to test them efficiently. Systolic arrays are not well suited to typical scan path techniques, however, which would attempt to gain access to the very large number of latches distributed throughout the array. Efficient testing ought instead to exploit the repetitive features of the array, and the theoretical concepts developed for testing iterative arrays of combinational logic cells are highly relevant. Key concepts are those of *C-testability* in which test sequences are propagated from one cell to another so that they are of constant length independent of the number of cells and *I-testability* in which the outputs of comparable cells are identical (Sridhar and Hayes, 1981).

We have recently presented a unifying framework for generating tests for fine-grained arrays (Marnane *et al.*, 1987, 1988, Marnane and Moore, 1988) but the purpose of this paper is to consider one particular example of a systolic array and explain in detail how it might be tested. Urquhart and Wood (1984) describe a bit-level systolic array which performs inner-product computations with a near 100% duty cycle. It has the architecture illustrated in Figure 9.1 for a small 4×4 array with the vector X flowing across the array from right to left in bit-serial, word-parallel form meeting bits of the other vector, W, held in the array. The array cells perform single bit multiplication by means of carry-save adders, and partial results flow down the array into an accumulator at the bottom. In applications such as convolution and filtering the vector W in the array may remain constant, but for more general applications a new value must be loaded for each calculation. This can be done in real time with the circuit shown in Figure 9.2 where the vector **W** is applied in word-serial, bit-parallel form along vertical data busses and the new values are loaded into the array at just the right instant as determined by a load control bit which is propagated with the 'wavefront' of the new calculation.

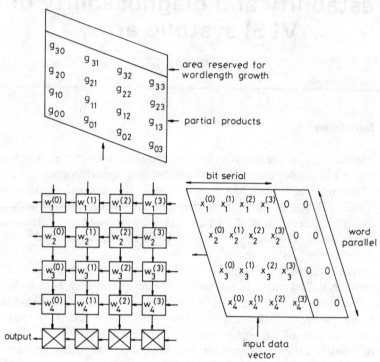

Fig. 9.1 *The systolic array of Urquhart and Wood (1984).*

9.2 Testability

9.2.1 *Testing the main array*

We describe first a strategy for testing the main array; the accumulator cells are considered separately later. Cells of the main array have the logic shown in Figure 9.3 and are described by the equations below. The design is based around half latches in the cells with alternate cells on alternate phases and Δ denotes the latch delay operator (i.e. a latch delay is half a clock period).

$$X' = \Delta (X)$$
$$S' = \Delta(S \oplus (X.W \oplus CTRL) \oplus C)$$
$$C' = \Delta(S.C. + S(X.W \oplus CTRL) + C(X.W. \oplus CTRL))$$
$$CTRL' = \Delta(CTRL)$$

This array has a number of particular features which we will exploit for testing.

(a) It is unilateral in that data only flow *down* the columns and *from right to left* along the rows.

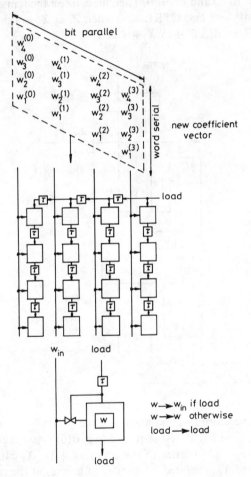

Fig. 9.2 *Loading **W**.*

(b) The row data paths are simple shift registers so that data pass along in this direction without being changed. This means that we can test $X' = \Delta(X)$ independently of the rest of the logic, and we can present adjacent columns with identical inputs in successive time periods. We can therefore test the columns in parallel and have identical results appearing from each column in successive time intervals. Apart from this propagation delay, the length of the test vectors is independent of the number of columns so that the array of columns is essentialy CI-testable (Sridhar and Hayes, 1981).

(c) The remaining logic in each cell can be split into two parts which can be tested separately. i.e. if we write $(X.W \oplus \text{CTRL}) = Z$, then we can fully test this bit-product term Z, with $S = C = 0$ when Z appears directly on the S output: $S' = \Delta(Z)$ and $C' = 0$. Then we can test the remaining logic (a full adder) with $W = 1$ and CTRL $= 0$ when $Z = X$, so that $S' = \Delta(S \oplus X \oplus C)$ and $C' = \Delta(S.C + S.X + C.X)$.

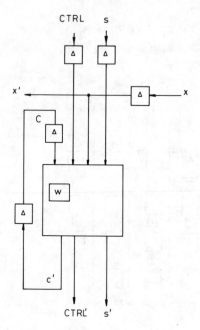

Fig. 9.3 *One cell of the array.*

The testing can therefore be conducted in three parts:

Test 1 (for the X paths): A test sequence (e.g. 010) is propagated along the X paths from the top right corner of the array. X_2, X_3, X_4 can most easily be delayed versions of X, generated by a series of latches at the right of the array as shown in Figure 9.4.

Results can be compressed by comparing appropriately delayed X outputs. The comparator might take the form suggested by Sridhar and Hayes (1981) of exclusive-OR gates comparing adjacent columns feeding into an OR-chain, but we prefer an AND-chain and an OR-chain as shown in Figure 9.5 which will differ from each other wherever there is an error. This is slightly simpler to implement and also gives a data output which can be checked for common mode errors.

Test 2: The Z terms ($Z = X.W \oplus \text{CTRL}$) can be tested by applying the eight combinations of X, W and CTRL to every cell. If Z is set to $Z = 0$ before and after these combinations and if the S data inputs at the top of the array are held

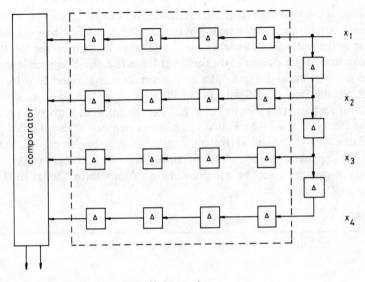

Fig. 9.4 *Arrangement for testing the X data paths.*

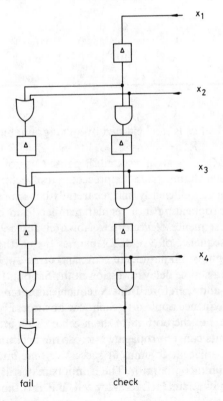

Fig. 9.5 *The row comparator.*

at $S = 0$ as they would normally be, the four values of $Z = 1$ which are generated at each cell propagate down the array and can be observed at the S' outputs at the bottom of the array. As CTRL and the load bit for the W data are propagated from the top right corner of the array, the X test sequence must also be propagated in the same way and this can once more be done by means of shift register latches as illustrated in Figure 9.4. A suitable sequence of test vectors for each cell is shown in Table 9.1. The 'wordlength growth' is necessary because values of $Z = 1$ generated in a cell can coincide with values of $S = 1$ passed down from the cell above and can therefore generate carries. Identical results then appear at the S' outputs at the bottom of each column at successive intervals of time and may be compressed by a comparator similar to the row comparator of Figure 9.5.

Table 9.1   *Suitable sequence of test vectors*

	Initialisation			8 test inputs								Word length growth		
CTRL	0	0	...	0	0	0	1	1	1	0	0	0	0	0	...
Win	×	0	...	1	1	1	1	1	1	0	0	0	0	0	
Load	1	0	...	0	0	0	0	1	0	0	0	0	0	0	
W	×	0	...	0	0	0	0	0	1	1	1	1	1	1	
X	×	×	...	×	1	0	1	0	1	0	1	0	0	0	
Z	×	0	...	0	0	0	1	1	0	1	1	0	0	0	

Test 3: The full adder cells can be tested by aplying suitable test vectors to the S and X inputs (with CTRL $= 0$ and $W = 1$ to give $Z = X$) such that the eight combinations of S, Z and C are generated. Because C is not directly accessible, it must be generated each time from the preceding test vector. This is the key to testing the systolic array efficiently and a carefully chosen sequence of vectors can result in tests propagating in a regular manner from cell to cell. In this example, a suitable sequence of vectors is shown in Table 9.2.

Notice how this sequence of S and X generates the next (recirculated) value of C such that all eight combinations of S, X, and C are applied, and also such that the S' output is a simple delayed version of the S input. In fact S' is delayed by 5Δ relative to S, and therefore if this S sequence is entered at the top of the column and the X sequence applied to each row at successive delays of 5Δ, the S vector will reappear at the bottom of the n-cell column after a delay of $5n\Delta$. Again W and S inputs can be propagated across the columns so that identical results appear from adjacent columns at successive time intervals and may be compressed in a column comparator. The X inputs may still be generated with a shift register of latches similar to Figure 9.4, but this time a delay of 5Δ is needed at each stage.

9.2.2 Combined tests

Tests 2 and 3 can be combined as shown in Table 9.3 in terms of the Z and S inputs and the generated values of C' and S' for the right-hand column of a 4×4 array. First the array in initialized by entering $(1 + \lceil \log_2 n \rceil)$* zeros to the S and Z inputs (via $W = \text{CTRL} = 0$); the $\lceil \log_2 n \rceil$ term here arises from the 'wordlength growth' of the carries. Test 2 can then be started; the initial vector takes eight periods and again can expand by $\lceil \log_2 n \rceil$ as it propagates through the cells. Test 1 for values of X can be conducted at the same time as the array is initialized and Test 2 performed. During initialisation, the X test sequence can be chosen arbitrarily because $W = \text{CTRL} = 0$ sets $Z = 0$ independently of X and during Test 2, X is given the sequence 10101010 anyway. Test 3 starts immediately after Test 2 because, despite the wordlength growth of Test 2, the effective values of Test 3 propagate down at only one cell every five half-periods. Test 3 takes $8 + 2(n - 1)$ periods plus propagation time. Finally tests on the other $(n - 1)$ columns take just another $(n - 1)$ half-periods to give a total test length of $(1 + \lceil \log_2 n \rceil) + 8 + (8 + 2(n - 1)) + n/2 + (n - 1)/2 = (3n + \lceil \log_2 n \rceil + 14.5)$ clock periods.

Table 9.2   *Suitable sequence of test vectors*

	8 test vectors											
$S =$	0	1	0	1	1	0	1	0	0	0	0	...
$X =$	0	1	1	1	0	0	0	0	1	0	0	...
which generate												
$C = \Delta C' =$	0	1	1	1	1	0	0	0	0	0	0	...
$\Delta S' =$	0	0	0	1	0	1	1	0	1	0	0	...

9.2.3 Testing the accumulator

The accumulator is another series of carry-save adders and can be tested with a similar pattern to Test 3 above. This time the equivalent of the controlling Z values must appear from the S' outputs along the bottom of the main array and this can be done by means of suitable W coefficients entered into the array and activated by $X = 1$ values entered at the bottom row first and on higher rows at successive intervals of Δ. The X values can still be generated by shift register latches as in Figure 9.4, but the delays, Δ, between rows must now be from the bottom of the array to the top. This test would take a further $(8 + 2(n - 1))$ periods, would take $(n - 1)$ periods to propagate through, and could follow straight after the main array tests to give a total test time of $(6n + \lceil \log_2 n \rceil + 19.5)$ clock periods.

---

* The ceiling function $\lceil u \rceil$ denotes the smallest integer greater than or equal to u.

Table 9.3 *Combined tests for a 4 × 4 array*

9.2.4 Siting of the test generators and comparators

In the design of a testable chip, consideration must be given to the placement of the test generators and the comparators. The sign control signal (CTRL) would normally be applied via a shift register anyway and we have seen that a similar generator can replicate the sequences of S test inputs required at the top of each column. The X values required for each row could also be replicated in this way, although the tests above require three different delays in such a generator. Comparators as in Figure 9.5 can be used to compare the X and S test outputs.

In general, if the signals come off chip it is better to place the generator (comparator) off chip so that all the I/O pins are tested and if the signals do not come off chip it is probably necessary to place the generator (comparator) on the chip to avoid using extra pins. For the row testing, all the X inputs would be available at the external pins so that the generator can easily be external to the chip. The X outputs may not be available externally and in this case the X comparator should be placed on the chip and will itself need to be tested. This will involve setting each row in turn to an erroneous 'zero' when the other rows are all 'one' and vice versa. For the column testing, access is required to the S inputs at the top of each column which would normally be set to 'zero' and would not be available externally. The S test sequence should therefore be supplied via a single external pin to the right hand column with an on-chip generator creating the delayed versions for the other columns. The S outputs at the bottom of the array would not normally be accessible either and so for test purposes should be taken to an on-chip comparator instead of (or as well as) the accumulator. Erroneous values to test this comparator cannot be entered via the on-chip S generator but can instead be entered via the W inputs.

Comparator testing would therefore take a further n clock periods to yield a final test time of $(7n + \lceil \log_2 n \rceil + 19.5)$ clock periods. For a typical 32×32 array of 1024 cells this would be just 249 clock periods.

9.3 Diagnosability

Systolic arrays can easily be designed to suit any size of problem and we have proposed the use of spare columns to obtain an adequate yield on large arrays (Moore *et al.*, 1986). Defective columns would be bypassed by the circuit shown in Figure 9.6. This requires that we can identify the columns which are free of defects and the pass/fail tests described above can quite easily be extended to this problem.

In principal the switching circuits of Figure 9.6 can allow each column to be tested individually. Test 1 could therefore be performed on each column taken one at a time. The accumulator cells can be tested at the same time providing that the column has been cleared and required a test sequence of eight clock periods plus a propagation time of one period. The X data paths along with the generator and comparator of Figures 9.4 and 9.5 would normally take the test vector length plus ($\lceil n/2 \rceil$ + 1) clock periods to test, but if this is greater than the nine clock periods of the accumulator test it could be reduced by applying the X sequence to all odd rows at the same time and then to all even rows on the next clock phase. The test time for all the columns will therefore be $9n'$ clock periods where n' is the total number of columns (i.e., $n' = n +$ spares). This approach requires long bypasses with potentially large numbers of pass gates in series, and it will then be necessary to include buffers in the bypass line at every column, or every few columns. Even so, these tests may have to be conducted at less than full speed to allow for the large number of buffers in series.

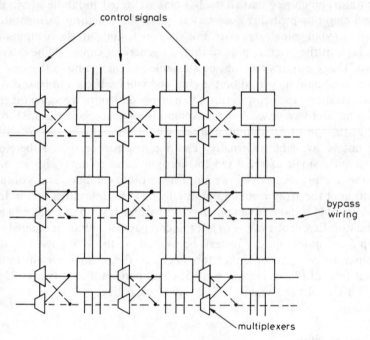

Fig. 9.6 *Column bypassing logic in the fault-tolerant chip.*

Tests could be devised to avoid these long delays; for example, if there are five spare columns it would be possible to test the array in six parts, each part formed by selecting one column out of every six. At least one of these parts must be fault-free if the chip is configurable and these columns can be enabled during a further round of tests with different combinations of the remaining rows enabled. After an appropriate series of tests a complete diagnosis of the chip could be made. Such a series of tests could easily be automated but is more complicated and would take longer than testing columns individually.

Another complication is when there are faults in the bypass lines. In the simpler test strategy described above a single bypass stage fault could be detected by the fact that the *only* test which works is the one where that column is enabled. Yield simulations for such a chip (Moore *et al.*, 1986) indicate that whilst it is useful to configure chips which contain bypass faults (indeed our choice of bypassing logic is a consequence of this) there will be a very small number of configurable chips with more than one bypass fault. If for a particular application this turns out not to be the case a more sophisticated test strategy would again be required. Some alternative strategies are discussed by Lewis and McCabe (1988).

Tests 2 and 3 can then be conducted as before on the reduced size array containing all the columns which pass Test 1. The results sequences from each column now need to be analysed separately and this can be done with the circuit shown in Figure 9.7 by comparing each output with the expected results sequence and latching any errors. A similar circuit is described by Cerny and Abdoulhamid (1983), but compares results with adjacent columns and is therefore inadequate for the diagnosis of multiple faults or common mode faults. The overall test time will therefore be $(1 + \lceil \log_2 n \rceil + 9n' + (8 + 2(n - 1)) + n/2 + (n' - 1)/2 + n') = (10.5n' + 2.5n + \lceil \log_2 n \rceil + 6.5)$. For a typical 32×40 array this would be 512 periods.

S' from array

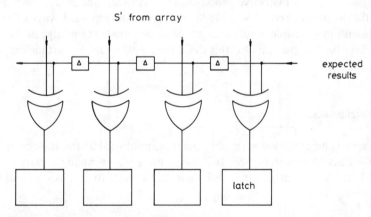

expected results

latch

Fig. 9.7 *Column comparator for the fault-tolerant chip.*

9.4 Conclusions

We have looked at the testing of a bit-level systolic array and have seen that it is possible to exploit its regularity to obtain simple test vectors, to compress results vectors and to obtain tests of length $O(n)$ for an $n \times n$ array. The tests described owe their origins to theoretical work on testing iterative logic arrays of combinational logic and are very much more efficient than scan path testing for a chip of similar complexity. On the other hand, where it is desirable to site generators or comparators on the chip, in order to reduce the test generation or results comparison, or to use signals which are not available at the pins, or because of a desire to make the chip self-testing, this can be done by means of strategically placed shift registers.

The approach is readily extended to the diagnosis of faulty columns as is required in a fault-tolerant version of the array. The test length is still of $O(n)$ and is typically about twice that of the original design. Complications can arise if multiple bypass circuits have too long a delay or if it is necessary to diagnose multiple faults in the bypassing circuits but these are not insurmountable.

The tests provide an exhaustive functional test of every distinct functional element of the array at full operational speed. Because of the spatial separation of the functional elements, the test will therefore give 100% coverage of all single and multiple permanent faults. We have not considered the problem of intermittent faults such as pattern sensitivity.

Thus what might have been a difficult circuit to test because of a large proportion of buried storage elements turns out to be testable with exceptionally simple tests and exceptionally high fault coverage. This further confirms the advantages of systolic arrays for VLSI designs. The full adder is a key element of a number of digital signal processing arrays (Urquhart and Wood, 1984) and all such arrays that we have examined to date have also turned out to be amenable to fast and effective propagating tests (Marnane *et al.*, 1987, 1988). Recently (Marnane and Moore, 1988) we have also reported ways of testing circuits with inaccessible inputs such as the carry signals at the edge of the array and it may be that the tests for this design example could be further simplified in this way.

Acknowledgements

This paper is an extended version of one first published at the European Solid State Circuits Conference 1985. It is based on work originally undertaken for GEC, Hirst Research Centre, and their permission to publish is gratefully acknowledged.

References

CERNY, E. and ABDOULHAMID, E. M. (1983): 'Built-in testing of Cl-testable iterative arrays', *Fault Tolerant Computing Symposium* pp. 33–6
LEWIS, A. and McCABE, A. P. H. (1988): 'A fault-tolerant test chip for gathering wafer defect data' in *Yield Modelling and Defect Tolerance in VLSI* (eds. W. R. Moore, W. Maly and A. J. Strojwas) (Adam Hilger, Bristol)
MARNANE, W. P., MOORE, W. R., YASSINE, H. M. GAUTRIN, E., BURGESS, N. and McCABE, A. P. H. (1987): 'Testing bit level systolic arrays', *Proc. Int. Test Conf.*IEEE pp. 906–14
MARNANE, W. P., MOORE, W. R and GAUTRIN, E. (1988): 'Structured test of bit level systolic arrays', in *Yield Modelling and Defect Tolerance in VLSI*, (eds. W. R. Moore, W. Maly and A. Strojwas, eds) (Adam Hilger, Bristol)

MARNANE, W. P. and MOORE, W. R. (1988) 'Testing of VLSI regular arrays,' *Int. Conf. on Computer Design* IEEE pp. 145–8

MOORE, W. R., McCABE, A. P. H. and BAWA V. (1986): 'Fault-tolerance in a large bit-level systolic array,' in *Wafer Scale Integration*, (eds. C. R. Jesshope and W. R. Moore) (Adam Hilger, Bristol)

MOORE, W. R., McCABE, A. P. H. and URQUHART, R. B. (eds.) (1987): '*Systolic Arrays*' (Adam Hilger, Bristol)

SRIDHAR, T. and HAYES, J. P. (1981): 'Design of easily testable bit-sliced systems', *IEEE Trans. Computers,* **C-30**, (11), pp. 842–54

URQUHART, R. B. and WOOD, D. (1984): 'Systolic matrix and vector multiplication methods for signal processing', *IEE Proc.,* **131** (Part F, No. 6), pp. 623–31

Chapter 10

Path testing of MOS circuits

R. I. Damper and N. Burgess

10.1 Introduction

Continuing advances in microelectronic integration, leading to ever increasing complexity of VLSI circuits and the imminent arrival of wafer-scale integration (WSI), offer great potential for more powerful yet cheaper products. However, increased functionality brings with it increased problems of testing. Any product needs to be tested to an extent that gives confidence it has been properly manufactured, but all testing is expensive and adds to product cost. Thus, such testing as is done should be maximally effective. This fact, together with the sheer difficulty of coping manually with the complexity of present and future VLSI and WSI circuits and systems, explains the increasingly important role of automatic test pattern generation (ATPG).

In this chapter, we first consider the possible approaches to automatic generation of test patterns for digital circuits. Although new materials such as gallium arsenide are starting to be used, the metal-oxide-silicon (MOS) technologies continue to dominate IC manufacture; for this reason, we concentrate on the problems of testing MOS circuits. The currently best-accepted approach makes assumptions about the way a circuit is likely to fail – so-called structural or fault-oriented testing – and tests for the presence (or absence) of such faults. This implies the existence of a *fault model*. In the past, the stuck-at model operating at a boolean-gate level of abstraction has proved its value in the testing of printed circuit boards (PCBs) populated with SSI and MSI components and, thus, has achieved a position of almost total dominance. We review evidence that the 'classical' stuck-at model is inappropriate for testing VLSI and WSI MOS circuits essentially because MOS processing faults do not generally lead to stuck logic nodes. Rather, they lead to the presence of conducting paths (between V_{DD} and V_{SS}) which should not exist under fault-free conditions, or to the absence of paths which should exist. Thus, the resulting fault effect is alteration of logic function rather than a stuck node. A better approach is to base structural testing on a *switch-level* model which has the capability to represent explicitly important aspects of MOS circuit operation. Thus, new test pattern generation (TPG) methods are required for use with such a model. We describe in some detail a method based on path algebras ideally suited to this purpose.

The path algebraic TPG method is considered in the context of small, combinational static and dynamic NMOS circuits ('primitives') as well as CMOS primitives. Subsequently, we show how the method can be used in conjunction with the well-established *D-algorithm* for the testing of larger assemblies. Path algebraic testing is then contrasted with other switch-level methodologies and, finally, we outline the implications of our approach for design practices.

10.2 Approaches to test pattern generation

Automatic testing of digital electronic systems typically consisting of TTL packages assembled on PCBs has become a relatively well-developed art (Bennetts, 1982), at least in the case of permanent faults. Although the detection of transient faults and increased delays remain a problem, it is nonetheless natural to attempt to apply the same proven methods to the testing of microsystems fabricated by newer process technologies. In PCB testing, two essentially complementary approaches have long been recognised. So-called *functional testing* applies stimuli to the input edge-connections of the PCB, measures the circuit's response at the edge-connections, and compares this to the known fault-free response. That is, access to internal circuit nodes is via 'primary' inputs and outputs only, a requirement which can cause problems of controllability and observability. In this case, test generation can only be achieved by having a rather full understanding of the functional relationship between the various components of the circuit. By contrast, *in-circuit* test uses a 'bed-of-nails' fixture to contact directly the internal nodes of the printed circuit so removing controllability and observability problems. By a process of 'backdriving', individual components can be momentarily isolated from the surrounding circuitry and tested without consideration for the function of the component in the complete circuit. This greatly reduces test programming complexity and cost relative to functional test. With highly-integrated VLSI or WSI circuits, lack of access to internal nodes effectively rules out the in-circuit method and functional test becomes mandatory. Improving access, and hence controllability and observability, is a major goal of design for testability (DFT).

Given that the functional approach is to be used, how does the test programmer go about producing a set of test patterns? For simple combinational digital circuits, testing can sometimes be achieved by applying all possible inputs and checking that the correct outputs results. Such *exhaustive* testing is clearly not possible in the case of large circuits or of sequential circuits where a specific output depends upon the ordering of the input stimuli. However, it may be a viable technique when employed within one of the DFT methodologies which partition the VLSI circuit into small combinational blocks (although pseudo-random test is more usually employed in this case.) Clearly, there is no particular difficulty in generating exhaustive test patterns.

Another possibility is to check that the circuit performs the function for which it was designed. Unfortunately, this is often referred to as 'functional' test, a term which is obviously confusable with the sense used above. Thus, we prefer the term *behavioural* test. Although superficially a well-motivated approach, since it is directly aimed at telling if the circuit 'works', some particular difficulties arise with behavioural testing. There is, of course, the difficulty of adequately specifying the designed function of a complex microsystem. A quantitative specification of test program quality in terms of fault cover is also difficult because the set of possible faults cannot be enumerated. Finally, since likely manufacturing defects are not considered, much testing time and effort can be expended in checking for physically unrealistic faults.

Hence, the currently-accepted best method of testing is to make assumptions about the ways in which the circuit-under-test can fail, based on knowledge of its structure, and then to test the circuit to check that none of these assumed faults is present. This is called *fault-oriented* (or 'structural') testing; and the assumptions made about modes of failure constitute the so-called fault model. One particular fault model which has been extensively used for PCB testing in the past is the *stuck-at model* (Eldred, 1959). For reasons given below, this model has become so entrenched that stuck-at testing continues to be generally used for VLSI. In the following section, we argue that a different fault model is required, operating at the transistor (or switch) level and capable of representing important aspects of MOS circuit behaviour.

10.3 Fault models for structural testing

10.3.1. Stuck-at model

Any fault model needs to satisfy two pre-requisites. First, it must abstract the complex relationship between fault and fault effect to its barest essentials, thereby simplifying the tasks of test pattern generation, evaluation etc. Second, the model must represent the underlying physical behaviour of the circuit under test in an adequately realistic fashion, so that high physical fault cover (as opposed to cover assessed against an enumeration of hypothetical faults) is obtained. To some extent these two pre-requisites are in contradiction as recognised by Hayes (1985) who writes:

> 'All the fault models proposed for digital circuits attempt to combine computational simplicity with an accurate modelling of real failures. The simplification or abstraction inherent in fault modelling means that some physical failures . . . will not be covered.'

Fault-oriented testing of digital circuits has a long history: in 1959, Eldred introduced the single stuck-at model to assist the testing of logic circuits. The underlying notion was that the dominant failure mechanisms would lead to one and only one logic variable being 'stuck' at either logical 1 (s-a-1) or logical 0 (s-a-0) rather than being controlled by input signals as in normal operation. Hence, the model operates effectively at the boolean gate level of description. Initially, it was applied to the testing of vacuum tube/diode circuits, but later was used very successfully for assembled PCBs populated with SSI/MSI (e.g. TTL) components.

The success of the model, and its widespread acceptability, was due to many factors. Certainly, it achieved simplicity while retaining a useful measure of physical realism for failures resulting from PCB assembly processes. For instance, a floating TTL input is indeed s-a-1 and a line connected to earth by a solder splash will be s-a-0. To a lesser extent, the model was also capable of representing realistically a number of bipolar processing faults internal to the TTL package (Beh *et al.*, 1982) although there are certain important faults for which this is not true (Damper and Burgess, 1989). The fact that the model operates at the boolean gate level of description was also entirely appropriate for TTL designs. Most importantly, however, it was found to perform well in practice in terms of simplifying test pattern generation yet producing adequately effective fault cover. For this reason, although the single-stuck-fault model could be extended to cover multiple faults and bridging faults (whereby adjacent nodes which should be separate are electrically connected and so adopt the same logic value), this has not usually been thought worthwhile (Wilkins, 1988).

In view of the model's familiarity and success in PCB testing, it is natural to attempt to retain it as the basis of testing VLSI circuits and systems fabricated using modern technologies. Indeed, even some very recent books on VLSI engineering (e.g., Dillinger, 1988) consider no other fault model. However, the adequacy of the physical fault coverage obtained by use of the stuck-at model in conjunction with a gate level description is currently a matter for debate.

10.3.2 The stuck-at model and MOS testing

An extensive body of evidence now exists documenting the major shortcomings of the stuck-at model when applied to MOS testing. Specifically, these are:

● The structural description of an MOS logic circuit at the boolean-gate level is not topologically equivalent to the circuit layout (Galiay *et al.*, 1980, Burgess *et al.*, 1984). This may result in the generation of tests for nodes that do not exist or failure to generate tests for nodes that do exist.

● The faults that occur in MOS integrated circuits tend to produce short or open circuits (Courtois, 1981, Burgess and Damper, 1984, Fantini and Morandi, 1985) and indeterminate (non-digital) logic levels (Banerjee and Abraham, 1983, Burgess *et al.*, 1984) rather than stuck nodes.

● MOS transistors allow bidirectional current flow – a characteristic not directly modelled on an equivalent gate-level representation (Galiay *et al.*, 1980, Burgess *et al.*, 1984). This property can produce fault-effects which are not predicted (and therefore not covered) by the stuck-at model.

● MOSFETs have a high OFF resistance and their inputs (gates) present purely capacitive loads, characteristics widely exploited in the design of dynamic (stored-charge) circuits. In the presence of certain faults, however, output nodes can act as parasitic latches (Wadsack, 1978) because the node is effectively floating and does not charge or discharge properly. This phenomenon is usually associated with CMOS circuits (under stuck-open conditions), but can equally well occur in pass transistor networks implemented in NMOS (Banerjee and Abraham, 1983, Chen *et al.*, 1984). According to Gai *et al.* (1983), stuck-open faults are of importance in all the 'high impedance technologies of MOS, NMOS and CMOS'. Parasitic latch behaviour causes severe testing problems since it leads to sequential operation, so test inputs must be properly ordered if the fault is to be uncovered (Reddy *et al.*, 1984).

In fact, the logical fault-effect of MOS failures is very often to alter the logic function implemented by the circuit, rather than merely to render one of the variables of that function constant or stuck. Thus, the classical, stuck-at model is deficient in failing to represent adequately the effect of important MOS faults, as neatly summarised by Chen *et al.* (1984):

'It is illuminating to state the key assumption underlying classical test generation – a faulty combinational circuit is still combinational and digital. For the example NMOS circuit, a simple s-a-0 fault produces sequential behaviour, and a s-a-1 fault produces analogue effects. This violation of the classical assumption is at the root of the problem.'

The obvious conclusion to draw is that a more realistic model is required, perhaps operating at a lower level of abstraction. However, this conclusion has been challenged on grounds which can be understood by recalling the essential purpose of fault modelling, namely to balance the two largely conflicting goals of accuracy and simplicity. Exactly where the balance lies is a subjective judgement and naturally those favouring the retention of the classical, stuck-at model for VLSI testing will argue that it achieves acceptable performance (in terms of uncovering defects) in concert with a high degree of simplicity. By contrast, modelling at a lower level is held to be overly complex.

On the issue of accuracy, a number of authors have attempted to show that classically-derived tests will cover many non-classical faults fortuitously – both in small circuits and in large assemblies of gates. For instance, Abraham and Fuchs (1986) analyse a 3-input NMOS NAND gate with gate-short faults to conclude:

'. . . if we consider whether tests generated for stuck-at faults will detect the transistor faults (even though they cannot be modelled as stuck-at faults), it can be seen that the stuck fault model is still viable.'

Considering larger assemblies, Hayes (1985) has expressed an influential point of view. Pointing out that empirical evidence exists to show that the stuck-at model provides acceptably good coverage of permanent physical faults, he advances the likely reason that near-exhaustive stimuli are being applied to particular gates in a large circuit as a 'side-effect of tests aimed at other gates in the same circuit'. As a result, 'most physical faults are likely to be revealed. . .' This analysis of the situation has recently been reiterated by Wilkins (1988).

We find such arguments aimed at retention of the classical methods to be intensely unappealing since they are based on implicit reliance in the side effects of stimuli intended for another purpose. Also, the force of argument rests on the applied patterns being nearly exhaustive which seems to de-emphasise the importance of test efficiency. In our view, it is preferable to direct test generation at uncovering substantive faults and this requires modelling at a more detailed level.

On the remaining issue of complexity, test engineers have traditionally focussed on the fact that higher levels of abstraction are inherently simpler and, as a consequence, have tended to believe that one should only reluctantly move to a more detailed model. For instance, Abraham and Fuchs (1986) state:

'A fault which is described at a very low level, for example at the level of transistors, may very accurately describe the physical phenomena causing the fault but, because of the extremely large number of transistors in a VLSI chip, the model may be intractable for the purpose of deriving tests . . .'

While essentially correct, there is a danger of over-stating this argument to the extent of retaining the higher-level model even when it is demonstrably unsuitable for capturing important characteristics of physical faults. In order to make use of existing, classical TPG programs, many workers have suggested various equivalent logic-level networks and extensions to model physical MOS faults (e.g., Wadsack, 1978, Roth *et al.*, 1984, Jain and Agrawal, 1985). It is entirely possible, however, that the resulting, modified model will actually turn out to be more complex than the lower-level model itself. For instance, Al-Arian and Agrawal (1987) suggest use of a 16-to-1 multiplexer with asynchronous feedback merely to model stuck-open faults. A further danger, and one that seems to have occurred in practice, is that the modification (being essentially *ad hoc* in nature) obscures understanding of certain of the underlying faults and leads to them being overlooked. We take up this theme again in §10.8.

In an attempt to gather factual evidence on the suitability of the gate-level stuck-at methodology for MOS testing, we subjected a number of chips of a medium-scale integrated circuit (16-to-1 multiplexer implemented as an NMOS pass-transistor network) to various differently-derived test sets (Burgess *et al.*,

1988). Two of the test sets used were directed at stuck-at faults in two different gate-level abstractions of the circuit – essentially sum-of-products and product-of-sums descriptions. (It should be noted that these are merely two possibilities among a theoretically infinite number of gate-level descriptions). A reference test set was designed manually to detect all MOSFET stuck-on or stuck-open faults and pinhole (interconnect-short) faults. It was found that one of the stuck-at sets correctly (but fortuitously) identified all the faulty chips whereas the other identified 12% of the faulty chips as being fault-free. Since there is no principled way of determining in advance which of the very many possible gate-level representations will correctly model real faults, we conclude that these cannot be used with any confidence for modelling MOS circuits; a lower level of abstraction is needed.

While circuit-level (analogue) models certainly have the ability to represent MOS faults faithfully, they are too computationally expensive to be generally useful for ATPG. Thus, although our view would not be universally accepted, we feel that test pattern generation for MOS VLSI/WSI should be routinely done at transistor-switch level. It is recognised that a penalty must be paid in terms of increased complexity but this is offset by enhanced fault coverage. In any event, standard hierarchical VLSI design methodology gives an elegant means of coping with this complexity – a point to which we return later.

10.3.3 Switch-level models

For some years now, the need for a level of abstraction intermediate between the boolean-gate and circuit levels to describe MOS circuits has been apparent. The switch-level model represents the circuit as a network of transistor switches connecting charge storage nodes. Electrical behaviour is modelled in a highly simplified way, typically using a small, discrete set of voltage levels and impedances (transistor 'strengths'), in conjunction with a multi-valued (e.g. ternary $\{0, 1, X\}$) algebra. Perhaps the most influential work in this area has been that of Bryant (1984) and Hayes (1986). Switch-level modelling allows important effects such as dynamic charge storage and MOSFET bidirectionally to be well represented in a way that the gate level abstraction cannot do – yet without involving the full complexity of the (analogue) circuit description. It has been applied to such tasks as logic and fault simulation, formal hardware verification, timing analysis and automatic test pattern generation (Bryant, 1987).

Thus, one solution to the problem of realistic portrayal of physical MOS faults would be to operate at the switch level, considering transistor stuck-open and stuck-on faults as well as open and short circuits in interconnections. Indeed, many authors have recently considered the problem of ATPG at the switch level (Chiang and Vranesic, 1982, Agrawal, 1984, Roth *et al.*, 1984, Reddy *et al.*, 1985). In this work, as a compromise between more realistic representation and computational tractability, the tendency is to use a simplified switch-level model in which differences in transistor strengths and in node capacitance values are ignored. As a consequence, logic conflicts and delay effects under fault conditions are not directly modelled. We adopt this approach also.

Automatic test pattern generation for MOS circuits can be viewed essentially as a path-finding problem. That is, physical faults lead either to the existence of 'sneak' conducting paths (e.g. between V_{DD} and V_{SS}, between input and output nodes of a pass transistor network) which should not exist under fault-free conditions, or to the absence of paths which should exist. Test patterns must be selected to detect the fault-induced presence or absence of such paths. If TPG is to be done algorithmically, a formal, mathematical framework is required and a branch of graph theory known as path algebra (Carré, 1979) is ideally suited for this.

Before dealing with the application of path algebras to the problem of generating test patterns, however, we first describe something about MOS circuit operation.

10.4 NMOS circuit operation

10.4.1 Fault-free operation

MOS logic circuits operate by switching current-flow paths from the positive power supply rail either to the negative rail or on to the input of another logic circuit, according to the gate-source potential differences of the MOSFETs. Such circuits are generally restricted to driver-load configurations or pass-transistor networks. For NMOS technology, we define a 'primitive' to be a single-load circuit and its associated driver network together with any pass-transistor network connecting its output to the input of a following primitive.

An MOS circuit can be represented by a labelled graph. The nodes of the graph are the internal circuit nodes, excluding primary inputs to the gate terminals of MOSFET's. The graph's edges correspond to the (bidirectional) interconnections between nodes, either via transistor source-drain connections or via interconnect. Edges are labelled, typically with a single letter, identifying either the primary input to the transistor's gate or the interconnect. The logical value of the label determines the state of conduction of the edge: under fault-free conditions, interconnect labels are always logically true.

Figure 10.1(*a*) shows a typical NMOS gate together with its graph representation (Fig. 10.1(*b*)) where *w* stands for 'wire'. For driver-load configurations, we adopt the convention of numbering the nodes from 0 to *n*, reserving 0 for V_{DD}, 1 for the output node and *n* for V_{SS}. Under fault-free conditions, the 0 to 1 pull-up connection is always conducting. Further, the condition where the load transistor is stuck-open has such a gross effect on operation that its detection can be guaranteed – see below. Hence, for all such circuits we condense the graph by coalescing nodes 0 and 1 and removing the loop *l* (Figure 10.1(*c*)). The logical function of the NMOS gate can be found by determining the set of elementary paths from node 1 to node *n* and taking the negation of their union. An elementary path is one visiting no node more than once. For the circuit of Figure 10.1, the set of elementary paths is {*ab, cd, awd, cwb*}, where *w* is identically true. If any one of these pull-down paths is enabled, the output will be logical 0. Conversely, to make the output logical 1, nodes 1 and *n* (= 4) must be separated by switching off the edges of a cut set. A proper (or minimal) cut set is a set of edges between two specified nodes such that if any one edge is switched on, the two nodes are no longer guaranteed separated. The set of proper cut sets between nodes 1 and 4, given that *w* cannot be switched off, is {*ac, bd*}.

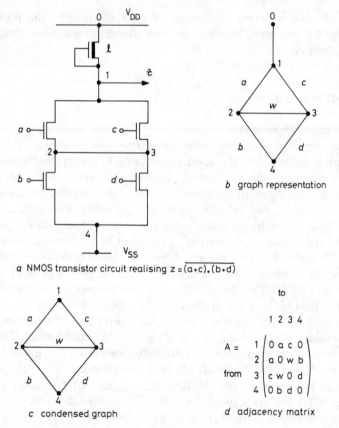

a NMOS transistor circuit realising $z = \overline{(a+c).(b+d)}$

b graph representation

c condensed graph

$$A = \begin{array}{c} \\ \text{from} \end{array} \begin{array}{c} 1 \\ 2 \\ 3 \\ 4 \end{array} \begin{array}{c} \text{to} \\ 1\ 2\ 3\ 4 \\ \left(\begin{array}{cccc} 0 & a & c & 0 \\ a & 0 & w & b \\ c & w & 0 & d \\ 0 & b & d & 0 \end{array} \right) \end{array}$$

d adjacency matrix

Fig. 10.1 *Circuit graph representation and adjacency matrix for simple NMOS digital circuit.*

The graph can be conveniently represented by the symmetric ($n \times n$) adjacency matrix, *A*, consisting of elements a_{ij} that take the label of the edge that connects node *i* to node *j* if one exists, or 0, the null element, otherwise. Figure 10.1(*d*) shows the adjacency matrix for the circuit under discussion. Note that *A* is a compact description of all the paths of length 1 in the circuit.

10.4.2   Operation under fault-conditions

With the switch-level model, fault effects are of two types – path-shorts and path-opens. Either conducting paths are produced which should not exist under fault-free conditions, or paths which should exist do not. Elementary paths have the property that, if any one edge is removed, the path becomes an open circuit. Thus, each path corresponds to a vector which tests for open circuits along that current path. Hence, the set {*ab, awd, cd, bwc*} for the graph of Figure 10.1(c) corresponds to the test vectors *abcd* = {1100, 1001, 0011, 0110} for the circuit of Figure 10.1(a), with labels included in the path assigned the value 1 and unspecified labels set to 0. All these vectors should make the output 0, but will fail to do so if any single transistor is stuck-open or *w* is open circuit. Analogously, the cut set {*ac, bd*} corresponds to vectors {0101, 1010} which produce a 1 output under fault-free conditions and test for any one transistor stuck-on or the depletion-load stuck-open. In this case, labels included in the cut are assigned the logical value 0 with unspecified labels set to 1. The usual techniques of fault collapsing are, of course, applicable. For instance, any three of the patterns {1100, 1001, 0011, 0110} will test for any transistor stuck-open and interconnect *w* present.

Thus, the set of elementary paths together with the set of proper cut sets, both determined for paths between nodes 1 and *n*, constitute tests for opens and shorts respectively under the usual single-fault assumption. In effect, they correspond to the singular cover for the *n*-node primitive gate. According to Roth (1966), the singular cover for a boolean-gate primitive:

> '. . . specifies *α*1, the totality of conditions under which the output of the block is 1, and *α*0, the totality of conditions under which the output is 0.'

The singular cover of a circuit element realising the logical function *f* is, in fact, a representation of the prime implicants of *f* and its negation *f̄* – the so-called *ON*- and *OFF*-arrays. The only difference here is that, whereas unspecified literals in prime implicants are assumed to be don't cares (*X*), in path testing they must be assigned the opposite logic polarity to the specified literals (as above). The analogy between prime implicants at the gate level and path and cut sets at the switch level has already been drawn by Chiang and Vranesic (1982), with TPG considered equivalent to the covering problem in logic synthesis.

TPG for the primitive gate is, therefore, equivalent to determination of appropriate path and cut sets. This we do by a novel method based on path algebras. In general, the relevance of path algebras to MOS design and test appears not to have been realised except that Hajj (1985) has (quite independently of our work) recognised their value in switch-level simulation.

Our development of the topic in §10.5 closely follows the exposition by Carré (1979).

10.5 Path algebras

10.5.1 *Fundamental concepts*

An algebra is defined as a set $\mathscr{P}$ equipped with a collection of *n*-ary operators. A path algebra is a set (having a zero and unit element) equipped with two binary operators, *dot* ($\cdot$) and *join* ($\vee$). The join operation is idempotent, associative and commutative while the dot operation is associative and distributive over $\vee$. The exact nature of the dot and join operators is determined by the path problem that the algebra is designed to solve. Since the singular cover involves both a path and a cut set, we require two algebras for ATPG purposes.

The algebraic solution of path problems can be understood by interpreting higher powers of the adjacency matrix, $\mathbf{A}^k$, as embodying information about edge-sets of order k. To find elementary paths, we require an algebra that produces $\mathbf{A}^k$ having elements a_{ij}^k describing paths of length k between nodes i and j. In ordinary matrix algebra, we find a_{ij}^k by computation of successive powers of $\mathbf{A}$ using multiply and add operations on the set of numbers. A path algebra generalises this procedure, replacing multiply and add by dot and join respectively, and operating on the set of edge labels. For instance, a_{13}^2 for the adjacency matrix of Figure 1(d) will be

$$\{0 \cdot c \vee a \cdot w \vee c \cdot 0 \vee 0 \cdot d\}$$

Similarly, length-2 paths between 1 and 4 will be given by

$$\{0 \cdot 0 \vee a \cdot b \vee c \cdot d \vee 0 \cdot 0\}$$

If we define the zero element such that $0 \cdot x = 0$ and $0 \vee x = x$, we can interpret dot as concatenation and join as union to yield:

$$a_{13}^2 = \{aw\} \quad \text{and} \quad a_{14}^2 = \{ab, cd\}$$

Should the join operation produce a path whose edges are a subset of another path, the set is reduced by removing the longer path. Hence, the operators for elementary-path determination are:

$$\text{dot: } a \cdot b \;\; = \;\; \{a \text{ concatenate } b\}$$

$$\text{join: } a \vee b \;\; = \;\; r\,\{a \text{ union } b\}$$

where r signifies reduce.

As no elementary path may visit a node more than once, every elementary cycle ($a_{i=j}$) and any path containing a repeated edge label (such as cca) is set to zero. This means no path may be more than $(n - 1)$ edges long and, as a consequence, we need evaluate powers of A no higher than $(n - 1)$. Figure 10.2 shows $\mathbf{A}$, $\mathbf{A}^2$ and $\mathbf{A}^3$ for the circuit of Figure 10.1 evaluated using this algebra. Thus, the required set of elementary paths between nodes 1 and $n\,(=4)$ is simply given by the join:

$$\bigvee_{k=1}^{n=1} a_{1n}^k = \bigvee_{k=1}^{3} a_{14}^k$$

$$= \{0, ab, cd, awd, cwb\}$$

The method relies on an element of the adjacency matrix satisfying the relation:

$$\bigvee_{k=1}^{q} a_{ij}^k = \bigvee_{k=1}^{q+1} a_{ij}^k$$

for non-negative integers q greater than some lower bound. An element is said to be stable if it satisfies this relation – known as the closure property. The least

$$\mathbf{A} = \begin{pmatrix} 0 & a & c & 0 \\ a & 0 & w & b \\ c & w & 0 & d \\ 0 & b & d & 0 \end{pmatrix}$$

$$\mathbf{A}^2 = \begin{pmatrix} 0 & cw & aw & ab,cd \\ wc & 0 & ac,bd & wd \\ wa & ca,db & 0 & wb \\ ba,dc & dw & bw & 0 \end{pmatrix}$$

$$\mathbf{A}^3 = \begin{pmatrix} 0 & cdb & abd & awd,cwb \\ bdc & 0 & 0 & acd \\ dba & 0 & 0 & cab \\ bwc,dwa & dca & bac & 0 \end{pmatrix}$$

Fig. 10.2 *Successive powers of adjacency matrix for Figure 10.1(d) computed using a path algebra for the determination of elementary paths.*

value of q for which closure holds is called the stability index. The same results and definitions can be applied to the adjacency matrix itself, when $\mathbf{A}$ is referred to as stable of stability index q. In the case under discussion, $\mathbf{A}$ is stable with stability index $(n-1)$, and the problem of finding elementary paths between output node and V_{SS} is equivalent to finding the closure of a_{1n}.

For cut set generation, a somewhat different algebra is needed giving the proper separating set of edges between nodes 1 and n. The required property of a_{ij}^k is that removal of the edges in the set should remove all paths of order k between nodes i and j. Considering paths of order 2 between nodes 1 and 4:

$$a_{14}^2 = \{0 \cdot 0 \vee a \cdot b \vee c \cdot d \vee 0 \cdot 0\}$$

To cut these paths, we can remove either edges ac, bc, ad or bd (Figure 10.1(c)) to leave paths of length 3 only. Thus, the required dot and join operators are:

$$\text{dot: } a \cdot b = r \{a \text{ union } b\}$$

$$\text{join: } a \vee b = r \{a \text{ concatenate } b\}$$

This gives:

$$a_{14}^2 = \{\{a, b\} \text{ concatenate } \{c, d\}\} = \{ac, bc, ad, bd\}$$

In this algebra, interconnect labels are first made equal to the unit element 1 (but see below), as these cannot be switched off. The unit element is defined such that $1 \cdot x = x$ and $1 \vee x = 1$. Again, we need only evaluate $\mathbf{A}^{n-1}$ since $(n = 1)$ (the stability index) is the maximum degree of node n. Figure 10.3 shows successive powers of $\mathbf{A}$ up to $\mathbf{A}^3$ computed using this algebra; note that $\mathbf{A}^k$ is symmetric. The required cut set is given by the join:

$$\bigvee_{k=1}^{n-1} a_{1n}^k = \bigvee_{k=1}^{3} a_{14}^k$$
$$= R\{0 \vee \{ac, ad, bc, bd\} \vee \{ab, ac, bd, cd\}$$
$$= \{ac, bd\}$$

where R denotes the reduce operation applied to a set of sets. As an illustration, to separate nodes 2 and 3 by disabling length-3 paths, we must remove all edges, $a_{23}^3 = abcd$ (Figure 10.3). This is necessary in order to destroy paths $aaw, bbw,$ wcc and wdd. (Note that path www remains: in this circuit a cut between nodes 2 and 3 is not possible. This fact could have been determined by leaving the label(s) for interconnect explicit i.e., not setting these labels equal to the unit element.)

Hence, using these two path algebras operating on the adjacency matrix, we can generate the singular cover for a primitive gate without difficulty.

10.5.2 Direct methods of solution

In the above illustration, all edges were considered to be bidirectional. However, edges connected to either power rail (via a load transistor where relevant) can be considered as unidirectional arcs to avoid determining current paths that traverse these nodes. This ensures that the circuit's graph is acyclic and guarantees that $\mathbf{A}$ is stable of index $(n - 1)$ under all path algebras.

For purposes of explanation above, higher powers of the adjacency matrix up to the stability index were calculated successively and their join formed. However, there are more efficient methods for determining matrix closure. Further, it is clearly unnecessary to evaluate elements of $\mathbf{A}^{(n-1)}$; only those elements contributing to a_{1n}^k need to be determined. Carré shows that the problem is formally similar to the classical problem of solving a system of linear algebraic

$$
\mathbf{A} = \begin{pmatrix} 0 & a & c & 0 \\ a & 0 & 1 & b \\ c & 1 & 0 & d \\ 0 & b & d & 0 \end{pmatrix}
$$

$$
\mathbf{A}^2 = \begin{pmatrix} ac & c & a & ac,ad,bc,bd \\ c & ab & ab,ad,bc,cd & d \\ a & ab,ad,bc,cd & cd & b \\ ab,ad,bc,bd & d & b & bd \end{pmatrix}
$$

$$
\mathbf{A}^3 = \begin{pmatrix} a,c & ab,ac,ad & ac,bc,cd & ab,ac,bd,cd \\ ab,ac,ad & ab,ad,bc,cd & abcd & ab,bc,bd \\ ac,bc,cd & abcd & ab,ad,cb,cd & ad,bd,cd \\ ab,ac,bd,cd & ab,bc,bd & ad,bd,cd & bd \end{pmatrix}
$$

Fig. 10.3 *Successive powers of adjacency matrix of Figure 10.1(d) computed using a path algebra for the determination of proper cut sets.*

equations. Hence, closure problems can be solved by variants of Gauss elimination, Jordan elimination, Gauss-Siedel iteration, etc. – the approaches differing from their classical counterparts only in the significance of the elementary operations. We have used the Gauss elimination method exclusively; this is guaranteed to find the required solutions (should they exist) since the graph of the circuit is absorptive (Carré, 1979) as a consequence of having no elementary cycles.

10.6 Pattern generation for MOS primitives

10.6.1 Static NMOS circuits
In §10.5.1 above, we showed how the two path algebras can be used to obtain the singular cover of a simple, static NMOS circuit (primitive) consisting of a

single depletion-load and its associated enhancement-driver network. The circuit in question had no MOSFETs in series or parallel and was irredundant. In Damper and Burgess (1987), we show that the circuit graph can be simplified by dotting the edges of transistors in series and joining the edges of transistors in parallel. Circuits with internal fan-out are tested by entering each transistor label separately in the adjancy matrix but with a numbered subscript identifying individual transistors controlled by the same input. Test vectors are generated in the usual way and the test sets then reduced, but treating $x_1, x_2, \ldots, x_n$ as the same label. If the label for a given transistor does not appear in the reduced set(s), it is logically redundant and produces no observable fault effect. To prevent redundant transistors from affecting fault detection, they should be switched off during test.

With the path algebraic methodology, it is straightforward to test for presumed short circuits within a primitive. Nodes between which short circuits may occur are simply specified by inserting a label identifying the suspected short circuit in the adjacency matrix. Before generating open-circuit tests in this case, labels for potential short circuits can be replaced by the zero element, 0. This ensures that no open-circuit tests are generated for a path that does not exist in the fault-free circuit.

10.6.2 Dynamic NMOS circuits

Dynamic circuits exploit the ability of the capacitive nodes of the MOS circuit to store charge for reasonable periods of time when in the 'high impedance' or disconnected state. Two principles commonly used in the design of dynamic circuits are charge steering and precharged logic. In charge steering, a pass transistor is used to transfer charge from one node to another under the control of the gate voltage which thereby acts as a clock. Precharged NMOS logic operates by unconditionally charging the output node from V_{DD} when the system clock is low (high) and conditionally discharging the node to V_{SS} when the clock is high (low) dependent upon the state of a driver network controlled by input logic values. In effect, precharged logic is a special case of charge steering in that charge transfer is always from/to one power supply rail to/from the output node.

NMOS designs based on charge steering suffer from the problem that stuck-open faults in the pass-transistor network result in alteration of logic function from combinational to sequential and, as a result, test patterns must be properly ordered. Since this fault is generally considered in the context of CMOS circuits, we outline our approach to the problem in §10.6.3 below.

According to Bryant (1987), very little work has been done on switch-level testing of dynamic circuits, and this is particularly so for the unconditional-precharge type. Generally, however, the effect of a stuck-open clocked transistor will be sufficiently catastrophic that its detection poses no problem. By contrast, a stuck-on clocked MOSFET means that precharging occurs under all clock conditions and this may or may not affect logical operation depending upon the relative ON resistance ratios of the various transistors in the path between V_{DD} and V_{SS}. Thus, unless special DFT design styles are adopted, untestable faults may arise. (One possibility is to monitor supply current for increases due to low resistance paths via the stuck-on MOSFET but this increase will need to be measured against the background of a relatively high drain by pad drivers and clock-distribution circuitry.) The remaining requirement is to test the conditional-discharge network driven by the inputs and here the situation is very similar to static NMOS testing so that the path algebraic method can be used to good effect.

10.6.3 CMOS circuits

A (static) CMOS primitive consists of a network of p-channel transistors between the output node and V_{DD}, and a complementary network of n-channel transistors between the output and V_{SS} sharing the same inputs. Under fault-free conditions, applying an input vector defines a current path between the circuit's output node and one of the power supply rails, and simultaneously switches off all such paths to the opposite supply rail. The output node is thus pulled either up or down as a function of which of the two transistors networks has a path activated through it. Because of complementary symmetry, the p-network path set and cut set are identical to the n-network cut set and path set respectively.

An important switch-level fault effect – that of stuck-open transistors – was first identified for CMOS circuits (Wadsack, 1978) and has received a good deal of attention in this context. (Indeed, many writers appear to believe, quite erroneously, that the stuck-open fault is peculiar to CMOS – e.g., Lala (1985), Dillinger (1988) – whereas it can occur in any of the MOS technologies.) As an illustration, suppose n-transistor *b* in the CMOS gate depicted in Figure 10.4 is stuck-open. On applying the input vector *abcde* = {11000}, a cut is applied to the p-transistor network. The corresponding path in the n-transistor network, however, does not pull the gate output low because the fault leaves it floating. In effect, cuts have been applied to both the n- and p-networks simultaneously. If the vectors {10010, 01100, 10001} were to be applied in that order (e.g. to test for all input s-a-0), the output would correctly remain at 0 for all three inputs and the fault would not be detected. Such faults can only be uncovered by a two-pattern test which first charges the output node and then attempts to discharge it through the transistor under test.

Care must be taken to ensure that no other discharge paths are temporarily activated during the transition from the first input vector to the second as a result of differences in transistor propagation delays (Agrawal, 1984). Tests conforming to this criterion are referred to as *robust* (Reddy *et al.*, 1984). As an example, consider the n-transistor network of Figure 10.4. If the two vectors were {00010} (to force the output to 1) followed by {11000}, n-transistor *a* may turn on before n-transistor *d* turns off, thus temporarily activating the path *ad* and discharging the output node but not through the transistor under test, *b*. Robust tests are guaranteed if the transistors in a cut set containing the tran-

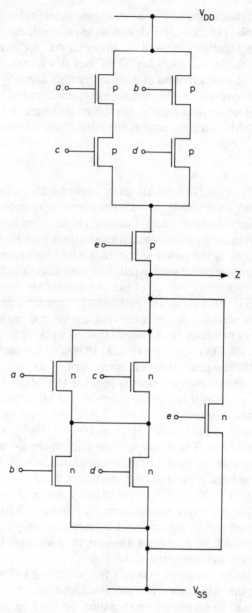

Fig. 10.4 *CMOS gate realising z = $\overline{(a + c + e) \cdot (b + d + e)}$.*

sistor under test remain off for the application of both input vectors. In this case, the output node must first be charged up by application of a cut set containing *b* and then discharged by apply a path set containing *b* but none of the transistors included in the initial cut. For example, the pair $\{X0X00, 11X00\}$

satisfies these conditions since all possible temporary paths must include *b*, the transistor under test. P-transistors may be tested for stuck-open faults if the tests are applied in the reverse order.

Newer dynamic CMOS techniques (e.g., 2- and 4-phase, domino) tend to remove this stuck-open problem since they replace the p-channel transistor network (implementing a conditional precharge and discharge scheme) with an unconditional precharge and conditional discharge scheme. As an illustration, Figure 10.5 shows the basic schematic for a domino-CMOS gate (Murphy and Edwards, 1981). Precharge is via a single clocked p-transistor, and there is a single series, clocked n-transistor in the path to V_{SS}. Thus, initialisation (charging of node X and discharging of node Y) takes place routinely between clock inputs. If either of the clocked transistors is stuck-open, the effect is catastrophic and easily detected; similarly, if either of the parallel p- or n-devices between X and Y is stuck-open. Overall, then, testing the n-device network driven by the primary inputs is the major problem and, again, path algebras are well suited to this. Indeed, the better testability of domino-CMOS compared to static CMOS has been confirmed by Oklobdzija and Kovijanic (1984).

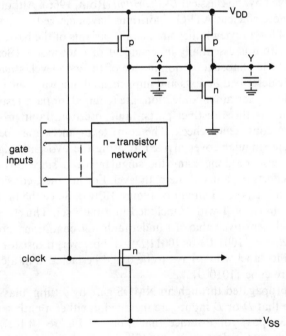

Fig. 10.5 *Schematic for domino CMOS gate.*

In the presence of a stuck-on fault, a CMOS gate acts as a voltage divider very much as in the case of dynamic NMOS. Hence, the gate output takes on an intermediate (non-digital) voltage which may be correctly regenerated by a following gate with the result that the fault is undetectable (Baschiera and

Courtois, 1984). However, the steady-state current drawn may well exceed electromigration limits leading to a low reliability component. One suggestion (Levi, 1981, Malaiya and Su, 1982) has been to monitor supply current for excessive leakage since the stuck-on device necessarily contributes to a low-resistance path across the power supply under certain test conditions. Because of CMOS's low quiescent current drain, this is more likely to be successful than in the case of dynamic NMOS (see above). One objection however, is that this form of monitoring will slow down functional test (Liu and McCluskey, 1987).

10.7 Path testing and the *D*-algorithm

10.7.1 Required modifications

So far, we have only considered test generation for primitive MOS gates in various technologies. Practical applications rely on generating tests for cascaded networks of primitives in which fault effects must be propagated to a primary output and the test stimuli traced back to primary inputs. This problem is essentially solved by the classical *D-algorithm* (Roth, 1966). Although a number of generally more efficient ATPG algorithms have emerged in recent years (see Bennetts, 1984 for a review), they are mainly variants of the basic *D*-algorithm.

The *D*-algorithm is essentially independent of fault model (Bennetts, 1982) but because of the almost total dominance of the gate-level, stuck-at model it is most usually described in texts and implemented in a way which is specific to that model. Hence, certain modifications are required for path-testing purposes. The algorithm has three distinct parts: fault insertion, fault propagation (or *D*-drive) and a consistency check. The consistency check may be taken unaltered, using the singular cover arrived at using the two path algebras. Fault insertion may also be done using the singular cover, since this conveniently describes the effects of faults at the path level. For instance, consider transistor *d* stuck-open in the NMOS circuit of Figure 10.1(a). Since the fault is an open, *d*'s gate input must be 1 with *d* included in a path set. Thus, assuming D to represent a fault-sensitive value of 1 under fault-free conditions, possible failure cubes are $abcd|z = \{1001|\bar{D}\}$ or $\{0011|\bar{D}\}$. If, however, transistor *d* were to be stuck-on then its label would have to be 0 and be included in a cut giving the required failure cube $\{1010|D\}$.

Faults are propagated through an NMOS gate by setting unassigned inputs to 1 or 0 such that D or $\bar{D}$ inputs are included in either a path set or a cut set respectively. Suppose for instance that input *a* (Figure 10.1(a)) is D under fault-free conditions. Selecting a path which includes *a* such as *ab* gives the propagating cube $\{D100|\bar{D}\}$. If, however, fault-sensitive input *a* had been $\bar{D}$, it would have been necessary to select the only cut set including *a*, namely *ac*, to give the propagating cube $\{\bar{D}101|D\}$. (Note the important fact that in this case the dual cube obtained by replacing D with $\bar{D}$ and vice versa is not also a propagating cube as it normally is.) Blocking or non-propagation *D*-cubes, whose appropriate use can greatly facilitate test generation (see Bennetts, 1982), can be obtained by analogous reasoning i.e. excluding inputs at D or $\bar{D}$ from any cut or path applied to the circuit.

An alternative method of determining propagating D-cubes would be to include inputs at D or $\bar{D}$ in both a path and a cut. For example, for input a fault sensitive, path ab and cut ac together give cube $\{D10X|\bar{D}\}$. Note that this cube could also have been derived by taking an intersection of the two propagating cubes in the previous paragraph, and that the dual cube $\{\bar{D}10X|D\}$ is now also a propagating cube.

Consider next CMOS primitives. Stuck-on faults may be inserted in the same way as for NMOS circuits, except that fault detection will perhaps be achieved by current monitoring (see above) rather than by propagating a fault effect to a primary output. With stuck-open faults, these usually but not always lead to sequential behaviour requiring a two-pattern test. For example, if n-transistor d (Figure 10.6) is stuck-open, all paths to ground are disabled and any path

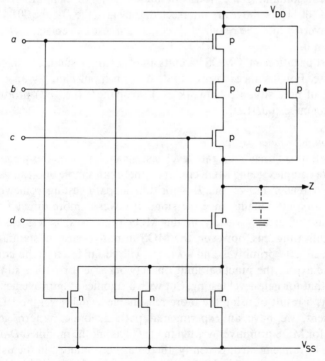

Fig. 10.6 *CMOS gate realising* $z = \overline{(a + b + c) \cdot d}$.

applied to the circuit will uncover the fault. Thus, only a single-vector test is required. Stuck-open transistors detectable by single tests correspond to cut sets consisting of a solitary transistor label. In the example, only n-transistor d falls into this category.

Insertion of sequential stuck-open faults is somewhat more complicated. For example, suppose we wish to postulate the fault n-transistor c stuck-open. In order to charge the output to $z = 1$, we apply an n-network cut including the transistor under test, abc say, with all other transistors not in the set (d) held at the opposite logic value. In this example, we arrive at the cube $abcd|z = \{0001|1\}$. We then invert the input to the transistor under test only to activate the n-path cd, whereupon output node z will fail to discharge. Thus, at this stage a failure cube $\{0011|\bar{D}\}$ has been produced. In the same way, to insert the fault p-transistor c stuck-open we apply a p-network cut including c (cd) to ensure $z = 0$ giving the cube $\{0011|0\}$. We then invert c whereupon z will fail to charge up so producing the failure cube $\{0001|D\}$. Note that the pattern $\{0011\}$ is common to these two tests so that the three-pattern test $\{0001, 0011, 0001\}$ can be conveniently used. Under fault-free conditions, the output z will then toggle from 1 to 0 and back to 1.

Alternatively, these failure cubes may be derived by intersection of both a path and cut containing the transistor under test. For example, if the cubes $\{0001\}$ and $\{0011\}$ as above are intersected, they give the D-cube $00\bar{D}1|D$ which is then converted into the required three-pattern test by replacing D by first 0, then 1, then 0.

Fault propagation in CMOS circuits is effected in similar fashion to the NMOS case. For instance, to propagate a D on input c, apply an n-network path (and, of necessity, a p-network cut) involving c (i.e., cd) producing the propagating cube $\{00D1|\bar{D}\}$.

10.7.2 Computational aspects

One important difference to gate-level testing is that here the primitives are rather more complex, being MOS circuits rather than simple boolean gates. This will tend to render use of the D-algorithm in path testing somewhat more expensive computationally since the singular cover is more difficult to derive and, because of greater variety in the MOS primitives, is less easy to pre-compile. Offsetting this, however, the MOS primitives are substantially larger than boolean gate primitives and so there will be far fewer to be considered. Overall, we expect the practical limit on network size for path testing will be similar to that for gate-level testing, but with a significant improvement in fault coverage as a result of using the more reasonable switch level model.

At present, we have an experimental path-algebraic pattern generation algorithm for MOS primitives written in DEC Pascal: the modified D-algorithm remains to be implemented. Recently, there has been interest in the use of logic programming languages for ATPG (e.g., Svanæs and Aas, 1984, Varma and Tohma, 1988), specifically because of their ability to handle the search and backtracking aspects of the problem, and we are currently considering a Prolog implementation of the modified D-altorithm.

10.8   Relation to other switch-level methodologies

In this section, we discuss path-algebraic testing in relation to other switch-level testing methodologies described in the literature.

Roth *et al.* (1984) recognise that testing based on the gate-level representation, without regard for the MOSFET implementation, has proven inadequate. They present a failure preserving switching- to logic-network transformation designed to retain important switch-fault information. Thus, tests generated to cover stuck-at faults in the so-called 'image logic' circuit using classical methods also constitute tests for transistor stuck-on and stuck-upon faults in the physical circuit. However, interconnection faults are not considered although it is clear that they could be. More seriously, memory (charge storage) effects are ignored. However, Jain and Agrawal (1985) (following Wadsack, 1978) specifically address this problem by adding a memory element to the set of logic primitives to model charge-storage fault effects. Consequently, they include an initialisation routine in the consistency part of the *D*-algorithm, thereby increasing computational complexity. Unfortunately, they also omit to consider interconnect failures which can lead to alteration of logic function. We believe that use of boolean logic primitives to describe an MOS switching circuit is potentially inefficient. Use of a more appropriate and perspicuous representation, such as path and cut sets, avoids any necessity to introduce spurious components to model real circuit effects.

A number of other methodologies basing switch-level testing on the *D*-algorithm are noteworthy. Reddy *et al.* (1985) show how the algorithm can be modified by considering transistor 'groups' (comparable to our primitives) in place of classical logic gates and Chen *et al.* (1984) also describe an extension to the *D*-algorithm for switching networks exhibiting non-classical faults.

Chiang and Vranesic (1982) adopt a radically different approach in using graph-theoretic methods exclusively without recourse to the *D*-algorithm. Their path-oriented representation explicitly models interconnection faults and MOSFET bidirectionally but the algorithm used is capable of finding cut sets only. Thus, to generate open-circuit tests, they find the cut set of the dual of the graph representation. This approach is restricted to circuits having planar graphs (since only such graphs possess duals.) Multi-level interconnect technology, however, makes it perfectly possible to implement circuits having non-planar graphs. In any event, it is difficult to determine computationally the dual of a graph by manipulation of its matrix description. Further, the method for generating tests at the higher (logic block) level is extremely cumbersome, as a consequence of using a graph representation at this level in preference to a *D*-algorithm formalism. Like us, El-Ziq and Su (1982) use a matrix-manipulation method to derive tests, but do not model MOSFET bidirectionality and do not guarantee to generate tests for interconnect faults.

10.9 Path testing and design for testability

Clearly, the path-testing methodology is based on knowledge of the circuit structure and, accordingly, test generation is most sensibly dealt with as part of the design process. Increasingly with VLSI and WSI design, hierarchical cell-based methods are seen as the way to cope with the complexity of the task and this applies as much to ATPG as to any other aspect of the design activity. Thus, we believe that existing methodologies offer a powerful means of dealing with any additional complexity resulting from use of switch-level (as opposed to gate-level) testing. The ultimate aim, therefore, should be to integrate automatic path test generation with other design tools.

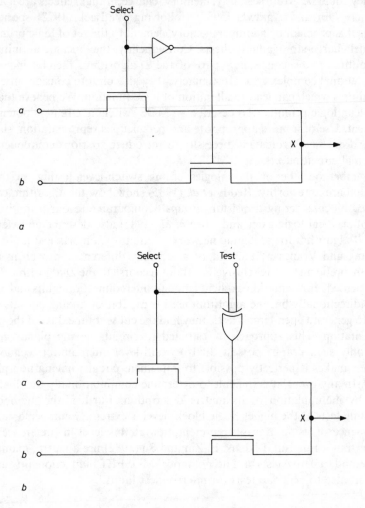

Fig. 10.7   *(a) Selector circuit without cut set, (b) selector circuit modified to be path-testable.*

In common with most present ATPG methods, path testing is restricted to combinational logic. The accepted way of dealing with this difficulty is to employ scan-test DFT techniques (Williams and Parker, 1982) which dynamically reconfigure the circuit under test into relatively small combinational blocks connected by shift-register chains.

There is, however, a class of MOS circuits using pass-transistor multiplexers that is not amenable to path testing because such circuits do not have cut sets. Hence, stuck-on transistors cannot be detected by our method. Such circuits are commonly used, and include the ubiquitous 'function block' (Mead and Conway, 1980) and selector circuits to resolve bus contentions. Their basic characteristic is that one logical selection variable switches off a subset of the pass transistors feeding a common node while its logical complement switches off the remainder. Figure 10.7(a) shows a simple illustration in NMOS technology. Although this problem could be dealt with using a more complex switch-level model in which signal strengths were represented, the probability remains that the fault effect would be an indeterminate logic level at node X. Fortunately, a simple design-rule alteration can ensure that it is always possible to produce a cut. All that is required is to replace inverters in the select lines by exclusive-OR gates, one of whose inputs is a test input (Figure 10.7(b)). If *test* = 1, the gate functions as an inverter; if it is 0, applying a 0 on any select line applies a cut across the transistors controlled by that line. Satisfying this design rule requires an extra $5n$ transistors for an NMOS circuit with n select lines or $4n$ transistors for CMOS technology.

With this rule observed, we consider path testing to be well suited to the requirements and constraints of the scan-test DFT methodology.

10.10 Summary

Structural testing of VLSI and WSI circuits in general requires non-classical fault models. For MOS technology certainly, and for others possibly, this model should operate at the switch level, intermediate between the classical (boolean) gate and analogue circuit levels of description. A unified and powerful method of test pattern generation for MOS circuits has been described, based on application of two path algebras to derive the path set and cut set of the circuit's graph.

Switch-level testing requires knowledge of circuit structure and, hence, test generation is properly a task for the design team. To make circuits path-testable, some design for testability constraints have to be accepted. One such constraint has been described in the case of selector circuits which, in their basic form, do not have usable cutsets.

References

ABRAHAM, J. A. and FUCHS, W. K. (1986): 'Fault and error models for VLSI', *Proc. IEEE*, **74**, pp 639–54

AGRAWAL, P. (1984): 'Test generation at switch level,' *Proc. IEEE Int. Conf. on CAD*, pp. 128–30

AL-ARIAN, S. A. and AGRAWAL, D. P. (1987): 'Physical failures and fault models of CMOS circuits,' *IEEE Trans. Circuits and Syst.*, **CAS-34**, pp. 269–79

BANERJEE, P. and ABRAHAM, J. A. (1984): 'Characterization and testing of physical failures in MOS logic circuits,' *IEEE, Design and Test*, Vol 1, No 3, pp. 76–86 (Aug.)

BASCHIERA, D. and COURTOIS, B. (1984): 'Testing CMOS: a challenge.' *VLSI Systems Design*, Oct. 1984, pp. 58–62.

BEH, C. C., ARYA, K. H., RADKE, C. E. and TORKU, K. E. (1982): 'Do stuck fault models reflect manufacturing defects?' *Proc. IEEE Test Conf.*, pp. 35–42

BENNETTS, R. G. (1982) *'Introduction to Digital Board Testing'* (Crane Russack, New York)

BENNETTS, R. G. (1984): *'Design of Testable Logic Circuits'* (Addison-Wesley, London)

BRYANT, R. E. (1984): 'A switch-level model and simulator for MOS digital circuits,' *IEEE Trans. Computers*, **C-33**, pp. 160–77

BRYANT, R. E. (1987): 'A survey of switch-level algorithms,' *IEEE Design and Test*, Vol 4 No 4, pp. 26–40 (Aug.)

BURGESS, N. and DAMPER, R. I. (1984): 'The inadequacy of the stuck-at fault model for testing MOS LSI circuits: a review of MOS failure mechanisms and some implications for computer-aided design and test of MOS LSI circuits,' *Software and Microsystems*, 3, pp. 30–6

BURGESS, N., DAMPER, R. I., SHAW, S. J. and WILKINS, D. R. J. (1984): 'Faults and fault effects in NMOS circuits: impact on design for testability,' *IEE Proc. (Part G)*, **134**, pp. 82–9

BURGESS, N., DAMPER, R. I., TOTTON, K. A. and SHAW, S. J. (1988): 'Physical faults in MOS circuits and their coverage by different fault models', *IEE Proc. (Part E)*, **135**, pp. 1–9

CARRÉ, B. A. (1979): *'Graphs and Networks'* (Oxford University Press, Oxford)

CHEN, H. H., MATHEWS, R. G. and NEWKIRK, J. A. (1984): 'Test generation for MOS circuits' *Proc. IEEE Test Conf.*, pp. 70–9

CHIANG, K-W. and VRANESIC, Z. G. (1982): 'Test generation for MOS complex gate networks,' *Proc. 12th IEEE Fault-Tolerant Computing Symposium*, pp. 149–57

COURTOIS, B. (1981): 'Failure mechanisms, fault hypotheses, and analytical testing of MOS LSI circuits' in *VLSI '81*, ed. J. P. Gray (Academic Press, London) pp. 341–5

DAMPER, R. I. and BURGESS, N. (1987) 'MOS test pattern generation using path algebras,' *IEEE Trans. Computers*, **C-36**, pp. 1123–8

DAMPER, R. I. and BURGESS, N. (1989): 'Faults, fault effects and fault models for VLSI' in Testability: Improving VLSI Quality, eds. A.P. Ambler, K. Baker and G. Musgrave (Prentice-Hall, Englewood Cliffs, N.J.)

DILINGER, T. E. (1988): *'VLSI Engineering'* (Prentice-Hall, Englewood Cliffs, NJ)

EL-ZIQ, Y. and SU, S. Y. H. (1982): 'Fault diagnosis of MOS combinational networks,' *IEEE Trans. Computers*, **C-31**, pp. 129–39

ELDRED, R. D. (1959): 'Test routines based on symbolic logic statements,' *J. Assoc. Comput. Mach.*, **6**, pp. 33–6

FANTINI, F. and MORANDI, C. (1985): 'Failure modes and mechanisms for VLSI IC's – a review,' *IEE Proc., (Part G)*,**132**, pp. 74–81

GAI, S., MEZZALAMA, M. and PRINETTO, P. (1983): 'A review of fault models for LSI/VLSI devices,' *Software and Microsystems*, 2, pp. 44–53

GALIAY, J., CROUZET, Y. and VERGNIAULT, M. (1980): 'Physical versus logical fault models,' *IEEE Trans. Computers*, **C-29**, pp. 527–31

HAJJ, I. N. (1985): 'A path algebra for switch-level simulation' *Proc. IEEE Int. Conf. on CAD*, pp. 153–5

HAYES, J. P. (1985): 'Fault modeling: tutorial,' *IEEE Design and Test*, Vol 2 No 2, pp. 88–95 (April)

HAYES, J. P. (1986): 'Psuedo-boolean logic circuits,' *IEEE Trans. Computers*, C-35, pp. 602–12

JAIN, S. K. and AGRAWAL, V. D. (1985): 'Modeling and test generation algorithms for MOS circuits,' *IEEE Trans. Computers*, C-34, pp. 426–33

LALA, P. K. (1985): *'Fault Tolerant and Fault Testable Hardware Design'* (Prentice-Hall, Englewood Cliffs, NJ)

LEVI, M. (1981): 'CMOS is more testable,' *Proc. IEEE Test Conf.*, pp. 217–20

LIU, D. L. and McCLUSKEY, E. J. (1987): 'Designing CMOS circuits for switch-level testability,' *IEEE Design and Test*, Aug., vol 4 no 4, pp. 42–9

MALAIYA, Y. K. and SU, S. Y. H. (1982): 'A new fault model and testing technique for CMOS devices' *Proc. IEEE Test Conf.*, pp. 25–34

MEAD, C. and CONWAY, L. (1980): *'Introduction to VLSI Systems'* (Addison-Wesley, Englewood Cliffs, NJ)

MURPHY, B. T. and EDWARDS, R. (1981): 'A CMOS 32-bit single chip microprocessor,' *Digest of Int. Solid State Circuits Conf.*, pp. 230–1

OKLOBDZIJA, V. G. and KOVIJANIC, P. G. (1984): 'On testability of CMOS-domino logic,' *Proc. IEEE Test Conf.*, pp. 50–5

REDDY, M. K., REDDY, S. M. and AGRAWAL, P. (1985): 'Transistor level test generation for MOS circuits,' *Proc. 22nd Design Automation Conf.*, pp. 825–28

REDDY, S. M., REDDY, M. K. and AGRAWAL, V. D. (1984): ' Robust tests for stuck-open faults in CMOS combinational logic circuits,' *Proc. 14th IEEE Fault-Tolerant Computing Symp.*, pp. 44–9

ROTH, J. P. (1966): 'Diagnosis of automata failures: a calculus and a method,' *IBM J. Res. Dev.*, 10, pp. 278–81

ROTH, J. P., OKLOBDZIJA, V. G. and BEETEM, J. F. (1984): 'Test generation for FET switching circuits,' *Proc. IEEE Test Conf.*, pp. 59–62

SVANÆS, D. and AAS, E. J. (1984): 'Test generation through logic programming,' *Integration: the VLSI Journal*, 2, pp. 49–67

VARMA, P. and TOHMA, Y. (1988): 'A knowledge-based test generator for standard cell and iterative array logic circuits,' *IEEE J. Solid-State Circ.*, JSSC-23, pp. 428–36

WADSACK, R. L. (1978): 'Fault modeling and logic simulation of CMOS and MOS integrated circuits,' *Bell Syst. Tech. J.*, 57, pp. 1449–74

WILKINS, B. R. (1988): 'Testing methodology: implications for the circuit designer,' *Microprocessors and Microsystems*, Vol 12, 1988, pp. 573–84

WILLIAMS, T. W. and PARKER, K. P. (1982): 'Design for testability – a survey,' *IEEE Trans. Computers*, C-31, pp. 2–15

Testing of mixed analogue and digital integrated circuits using the macromodelling approach

J. B. Hibbert, A. P. Dorey and P. J. Silvester

11.1 Introduction

An increasing proportion of the market for integrated circuits consists of custom designed ICs, which are not designed at the transistor level but are based on libraries of digital and analogue macrocircuits which may be interconnected to form the final system.

Test pattern generation (TPG) procedures for solely digital ICs are already well established, but mixed analogue and digital circuits pose new problems of testability.

The simulation and TPG for the analogue and digital parts of a design have usually been performed separately, and the handling of signals at the interface between them is often done manually. This chapter describes a method of integrating the analogue and digital test procedures through the use of behavioural simulation.

Clearly any test strategies that are to be developed should take advantage of the current trend toward a limited number of general-purpose macrocircuits that can be used time and again in the design of an IC.

Since we know which macrocircuits are likely to be used in the design of an IC (including opamps, analogue to digital converters and logic gates, which will be provided as macrocells by the manufacturer) and we can predict the faults which are likely to occur within these cells, then we are able to select test vectors which are of most use to reveal the faults. By using this approach when a particular IC is designed, test vectors for fault detection may be generated rapidly to provide efficient test strategies for the complete circuit.

Table 11.1 shows a selection of manufacturers who are currently offering application specific ICs (ASICs) which include analogue functions based upon a library of macrocircuits. The most popular analogue macrocircuits are listed; in addition individual manufacturers are offering macrocircuit such as analogue multiplexers, Schmitt triggers, monostables and sample-and-holds. Also shown in Table 11.1 are the digital macrocircuits offered by some manufacturers on the

TABLE 11.1 *IC Manufacturers offering macrocircuit-based designs*

Company, product and technology	ADC	DAC	DP-AMP	Compa-rator	VCO	Voltage ref	Current ref	Current mirror	Analog switch	Level shift	SSI	MSI/LSI	RAM/ROM	LSI/VLSI
						Analogue macrocircuits						*Digital macrocircuits*		
EXAR 'FLEXAR' BIPOLAR	—	×	×	×	—	×	—	—	—	—	—	—	—	—
FERRANTI 'MACROCHIP' BIPOLAR	—	×	×	×	—	×	—	—	—	—	—	—	—	—
MIETEC 'MADE' CMOS 3u	×	×	×	—	—	×	×	×	×	×	×	×	×	—
MIKRON 'MIGATE' CMOS 3u	×	×	×	—	×	×	×	—	×	—	×	×	—	—
PLESSEY 'MEGACELL' CMOS 2u	—	×	×	×	—	—	—	—	—	—	×	×	×	×
RAYTHEON 'RLA120' BIPOLAR	—	—	×	×	—	—	—	—	—	—	—	—	—	—
ZYMOS 'ZYCOM II' CMOS 5u	×	×	×	×	×	×	×	×	×	×	×	×	×	×

same chip as the analogue functions. These digital macrocircuits may be cat-
agorised into four groups corresponding to levels of circuits integration, as
shown in Table 11.2.

Table 11.2   *Circuit integration*

Level of integration	Examples of macrocircuits
SSI	gates, flip-flops
MSI/LSI	counters, registers, multiplexers
RAM, ROM	memories in any organisation, PLAs
LSI/VLSI	ALUs, microprocessors

11.2   Test methods: fault dictionary

A common approach to testing an analogue circuit is to compile a fault dictionary, as described by Yin and Elcherif (1985). A low-level circuit simulator such
as SPICE is used to simulate the circuit to be tested, both under fault-free DC
conditions and with a number of singly-chosen hard faults (open or short
circuits).

The input vectors selected for the simulation are chosen by a test engineeer
so as to make the faults distinguishable at the circuit output. The output vectors
of all these simulations are stored as the fault dictionary, which is simply a table
of single faults, input vectors and their corresponding output vectors.

In practice more than one input vector is needed for each fault in order to
achieve adequate separation of faults. There is currently no automatic method
of selecting these input vectors, and a knowledge of the way in which the circuit
operates is necessary at this stage.

It is possible to compile a fault dictionary for digital circuits in the same
manner as for analogue circuits. This method of detecting and locating a fault
is not particularly efficient for digital ICs, for which there are several alternative
approaches to testing, as described by Wilkins (1986). However, the fault
dictionary approach to testing has the advantage that it can be applied to ICs
containing mixed digital and analogue circuitry.

The main drawback with the fault dictionary method of testing is that the
circuit simulation can become enormously time consuming: SPICE is aimed at
producing accurate predictions of circuit behaviour for relatively simple circuits,
and the simulation of IC-size circuits for every possible hard failure is impracticable in all cases of any complexity. The next section (§11.3) looks at macrosimulation as a means of achieving a very fast DC simulation with sufficient
accuracy to create a dictionary of hard faults.

An advantage of the fault dictionary approach to testing is that the simulation
of fault effects need only be done once for each circuit, after which the analysis
of faults in the circuit is relatively fast. This 'post-test analysis' consists of

applying the input vectors to the circuit and comparing the measured output vectors with those stored in the fault dictionary, and hence determining which fault, if any, is present.

When considering a system composed of analogue and digital macrocircuits the fault dictionary for the final system is constructed in two steps. Firstly the fault effects of an open or short circuit in each macrocircuit require the construction of a new macromodel for each fault. These 'fault macros' may be reduced in number by performing fault collapsing, which is the grouping together of those fault macros with identical or near-identical fault effects at the output. The second step, as suggested by Dorey *et al.* (1987) is to substitute these fault macros for the fault-free macros and hence simulate the effect of every fault within the complete circuit (Figure 11.1). A second stage of fault collapsing may then be performed prior to storing the results in a fault dictionary.

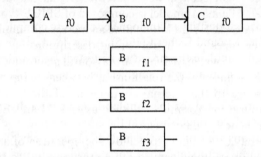

Fig. 11.1 *Substitution of fault macros ($f_1 - f_n$) or fault-free macros (f_0) to compute observable fault effects.*

Note that the writing of fault macros is a job that needs to be done only once, and the subsequent construction of a fault dictionary for a circuit is considerably speeded up if it has been designed using macrocircuits.

This method of devising test strategies assumes that only one fault is present in a circuit at any one time. If this 'single fault assumption' were not made then the number of possible collective faults in a circuit would become totally unmanageable. Hughes and McCluskey (1986) looked at the problem of multiple stuck-at faults in digital circuits and to what degree they are exposed or remain hidden when single stuck-at tests are used. Their analysis on a four-bit ALU, the 74LS181, showed that in simulated tests, all single stuck-at fault tests uncovered 99.96% of double stuck-at faults. Similarly Jha (1986) has shown that a test set which detects single stuck-open faults in any CMOS complex gate also detects all multiple stuck-open faults in it.

11.3 Macromodelling

Creating a macromodel of an electronic circuit consists of obtaining a model which will predict the output vectors from a macrocircuit in response to arbitrary vectors applied at its inputs. The model need bear no resemblance to the components used in the macrocircuit, and it should have minimal complexity whilst preserving sufficient accuracy so as to significantly increase the speed of simulation.

Macrosimulation consists of interconnecting such macromodels by defining a net list, preferably using schematic capture, to form a complete system and then running a suitable simulator to predict or verify the performance of the system.

There are many different forms of macromodel and six distinct types have been classified by Hsieh and Rabbat (1978). One of these types is the circuit reduction macromodel in which the original circuit is simplified. This has been demonstrated by Boyle *et al.* (1974) in the creation of a macromodel of a typical operational amplifier. In this method parts of the real circuit are modelled using simple ideal elements in such a way that the original circuit performance specification is met. The macromodel is composed of SPICE-type components, and when run on the SPICE simulator shows an increase in simulation speed of between six and ten times over the full circuit description.

Another method of macromodelling is behavioural macromodelling in which the macromodel is defined by the relationship between its inputs and outputs without any reference to the internal components of the macrocircuit being modelled. Behavioural macrosimulators such as HILO3 and ELLA are commonly used to simulate logic circuits.

These digital behavioural simulators allow the operation of a digital macro-circuit – anything from a logic gate to a microprocessor – to be described in textual form in a hardware description language (HDL). The HDL allows the signals of a digital macrocircuit to be defined as a function of:

(a) the input signals
(b) the current state of the macrocircuit
(c) delays associated with the macrocircuit.

Another behavioural simulator is HELIX, which is part of a CAD system available from Silvar-Lisco (1986). The HELIX HDL is aimed primarily at modelling digital circuits, but it also allows the descriptions of systems such as a chemical plant, and abstractions such as a project time schedule. The HELIX HDL is sufficiently flexible to allow the possibility of modelling analogue macrocircuits. This, together with the extensive libraries of commonly used digital circuits supplied with HELIX, make it an attractive choice to be used in building a DC fault dictionary for mixed analogue and digital circuits.

11.4 Digital macromodelling using HELIX

The behavioural simulator HELIX is part of the Silver Lisco SL2000 system of integrated CAD software. It provides multilevel simulation of a range of circuit abstractions, for instance software, architecture, register, functional and gate level. At the functional and gate levels, Silvar Lisco provide a number of model libraries comprising many TTL, CMOS and ECL logic functions. The designer uses the HIERARCHICAL HDL (HHDL) to customise and extend the model libraries. The HHDL modelling language is based on Pascal, with extensions to support concurrent modelling which effectively allow subprocesses to be executed simultaneously. It allows for synchronous and asynchronous circuits, user defined net-types, and may include standard Pascal mathematical functions. Net-type translation is also provided, using a model with zero delay that specifies how two nets of differing types will interact when connected together.

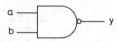

```
COMPTYPE nand;
INWARD a, b: BOOLNET;
OUTWARD y: BOOLNET;

SUBPROCESS
do_nand: TRANSMIT NOT (a AND b) CHECK a, b TO y DELAY 10;

BEGIN
(* Initialisation *)
END;
```

Fig. 11.2 *A NAND gate and its HHDL model*

A simple HELIX HHDL macromodel describing the operation of a NAND gate is shown in Figure 11.2. The input and output for this model can have a value of TRUE or FALSE: a two-valued model. The subprocesses 'do_nand' runs concurrently with any other subprocesses and is activated whenever the inputs *A* or *B* change value. This causes the value of NOT (*A* AND *B*) to be transmitted to the output Y with a delay of 10 ns. The 'BEGIN–END' part of the model is called the main body: it is executed prior to the start of the simulation, and is used to initialise the model, for example in setting up the contents of a ROM.

Simulation is performed by first encoding the HHDL behavioural models for all the components to be used. These models are processed by the HHDL compiler which checks the syntax, declarations, type compatibility and so forth. Next the circuit interconnections are entered using schematic entry. Once the behavioural and schematic information are available the designer uses the simulation linker to check the correspondence between the schematic and the behavioural descriptions, and to generate the simulator which is compiled, loaded and executed. The results of simulations may be output in tabular format or may be viewed using the Silvar Lisco logic analyser (LOGAN), which is used to display waveform output for logic circuits in graph form.

The HELIX HHDL is a very flexible modelling language but care must be taken in writing even the simplest of models, as it is easy to make mistakes, as shown in Figure 11.3(*a*). This HHDL model describes an inverter with a low-to-high propagation delay of 15 ns and a high-to-low propagation delay of 5 ns, and at first sight looks to be correct. However it does not take into the fact that an input pulse may be shorter than the propagation delays. Figure 11.3(*b*) shows the output waveform when a short pulse is applied to the input of this model. This user-written model should have included a check for a violation of the minimum input pulse width and should have issued an appropriate warning.

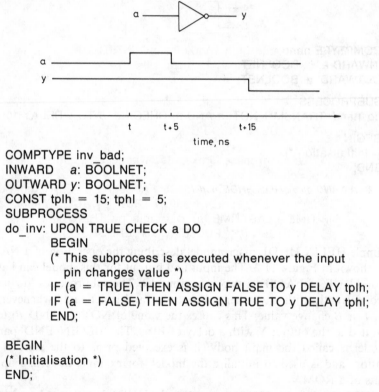

```
COMPTYPE inv_bad;
INWARD    a: BOOLNET;
OUTWARD y: BOOLNET;
CONST tplh = 15; tphl = 5;
SUBPROCESS
do_inv: UPON TRUE CHECK a DO
        BEGIN
        (* This subprocess is executed whenever the input
            pin changes value *)
        IF (a = TRUE) THEN ASSIGN FALSE TO y DELAY tplh;
        IF (a = FALSE) THEN ASSIGN TRUE TO y DELAY tphl;
        END;

BEGIN
(* Initialisation *)
END;
```

Fig. 11.3 *(a) An inverter and its incorrect HHDL model, (b) Waveforms for the HHDL model of an inverter*

The corrected model is shown in Figure 11.4. It illustrates the use of parameters, which may be passed from the schematic diagram (on a graphics terminal) to the HHDL model, enabling one model to represent components with differing time parameters, or any other defined parameter such as resistance, voltage, gain, etc. The three parameters for the inverter are the propagation delays (tplh, tphl) and the minimum pulse width (min_pw). In the case that no parameters are specified on the schematic diagram, default values are given within the model. To check that the minimum pulse width is exceeded two functions are called:

TIME: returns the current simulation time
LAST TIME (pin):   returns the time at which the previous
                    value of the pin came into being.

Each time the input 'a' changes value the output is inverted, after the appropriate propagation delay, and a check is made on the duration of the input pulse: if it is too short then a warning is written to the terminal during the simulation run.

```
COMPTYPE inv_good (tplh, tphl, min_pw: INTEGER);
DEFAULT tplh = 15; tphl = 5; min_pw = 10;
INWARD    a: BOOLNET;
OUTWARD y: BOOLNET;

SUBPROCESS
do_inv: UPON TRUE CHECK a DO
        BEGIN
        (* This subprocess is executed whenever the input
          pin changes value *)
        IF (a = TRUE) THEN ASSIGN FALSE TO y DELAY tplh;
        IF (a = FALSE) THEN ASSIGN TRUE TO y DELAY tphl;
        IF TIME – LASTTIME (a) < = min_pw THEN
            WRITELN ('*** PULSE_WIDTH ERROR ***');

        END;

BEGIN
(* Initialisation *)
END;
```

Fig. 11.4 *Corrected HHDL model of an inverter*

11.5 Analogue macromodelling using HELIX

In a similar manner it is possible to describe the behaviour of many analogue macrocircuits using the HELIX HHDL. A simple example is the Thèvenin equivalent circuit shown in Figure 11.5. The signal values are declared as REAL numbers to maintain accuracy in the analogue parts of a circuit, and the values of V and R for the Thèvenin circuit are passed to the model as parameters from the schematic diagram for convenience; default values must be specified in case the parameters are omitted. Note that there are no subprocesses in this mode, just the initialisation: the output signal is specified in terms of open-circuit voltage, and resistance to ground, and these values do not vary with time for a Thévenin equivalent source. Hence the main body of the model consists simply of asserting the values of V and R on the pin 'SIGNAL_OUT'.

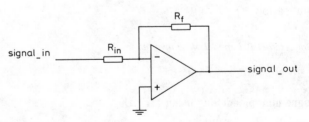

```
TYPE volts  : REAL;
      kohms: REAL;
NETTYPE signal-type = RECORD
                      V: volts;
                      R: kohms
                      END;
COMPTYPE Vsource (V: volts; R: kohms);
DEFAULT  V = 5.0;
         R = 1.0;
OUTWARD signal_out: signal_type;
VAR temp_signal: signal_type;

BEGIN     (* Initialisation *)
temp_signal.V: = V;
temp_signal.R: = R;
ASSIGN temp_signal TO signal_out;
END;
```

Fig. 11.5 *Symbol HHDL description of a Thevenin equivalent circuit*

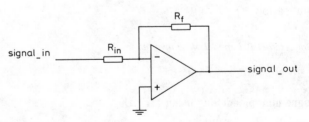

Fig. 11.6 *Opamp symbol.*

Other analogue macrocircuits may be modelled in similar fashion: output pins assert values of V and R, input pins accept values of V and R, while the main body and subprocesses of the model describe the relationship between these input and output variables.

Figure 11.6 shows the symbol for an opamp configured as a simple inverting voltage gain stage; the input and feedback resistors are included as part of the macrocircuit. The results of SPICE simulations and experimental results on the opamp show that it can be described, with reasonable accuracy, as having three linear operating regions, and that the breakpoints defining these regions are the two input voltages which cause the output to saturate.

A possible model for this in the HELIX HHDL is shown in the box below.

```
COMPTYPE opamp (Rin, Rf: kohms);
DEFAULT   Rin = 1.0; Rf = 1.0;
INWARD signal_in    : signal_type;
OUTWARD signal_out: signal_type;
CONST     supply_pos = 10.0; supply_neg = - 10.0;
          Vout_min = - 8.44; Vout_max = 8.94; V_offset = 0.003;
          Rout_1 = 0.83E - 3; Rout_2 = 0.8E - 4; -
          Rout_3 = 0.83E - 3;


VAR       temp_signal: signal_type;
          breakpoint_1, breakpoint_2: volts;


SUBPROCESS
input_change: UPON TRUE CHECK signal_in DO
BEGIN
IF signal_in.V < = breakpoint_1 THEN
    BEGIN
    temp_signal.V: = Vout_max;
    temp_signal.R: = Rout_1;
    END;
ELSE IF signal_in.V < = breakpoint_2 THEN
    BEGIN
    temp_signal.V: = -(signal_in.V*Rf)/Rin + V_offset;
    temp_signal.R: = Rout_2;
    END;
ELSE
    BEGIN
    temp_signal.V: = Vout_min;
    temp_signal.R: = Rout_3;
    END;
IF (signal_in.V > supply_pos)
OR (signal_in.V < supply_neg)
THEN WRITELN
    ('* * * * INPUT TO OPAMP IS ', signal_in.V, 'VOLTS * * * *');


ASSIGN temp_signal TO signal_out;
END;
```

```
BEGIN (* Initialisation *)
breakpoint_1: = - (Vout_max - V_offset) * Rin/Rf;
breakpoint_2: = - (Vout_min - V_offset) * Rin/Rf;
END;
```

The input and feedback resistor values (R_{in} and R_f) may be passed to the model as parameters from the schematic diagram, and default values are specified in the second line of the model. The CONST declaration defines eight constants: the two power supply voltages, the minimum and maximum output voltage swings, the output offset voltage, and the output resistances for each of the three operating regions. The subprocess 'input-change', when activated by a change in the input signal, first determines which region of operation is relevant for that input and then assigns the output values of V and R as appropriate. The initialisation section of the model calculates the values of the two breakpoints which define the linear operating regions of the opamp.

In order to demonstrate the speed advantages of a HELIX macromodel the above opamp model was stimulated in a circuit consisting of five interconnected opamps, the individual gains of which were selected to exercise the opamps over different operating regions. This circuit was simulated over the full input voltage range of the devices, and the results were compared with those of the same circuit using the equivalent Boyle *et al.* (1974) macromodels running under SPICE. It was found that over the full range of voltages applied (-10 V to $+10$ V) the output voltages agreed to within 10 mV which, whilst not being accurate compared to a full device-level simulation, is quite acceptable for rapidly predicting the effect of faults propagating through a large IC, and so for producing a fault dictionary. More importantly the circuit modelled using HELIX showed a typical increase in simulation speed of ten times that of the Boyle macromodel, which is itself estimated to run six to ten times faster than the device-level model.

11.6 Fault macromodelling using HELIX

To model a hard fault in an analogue macrocircuit is often easier than modelling the fault-free circuit, as open and short-circuits frequently result in the outputs being stuck at discrete voltage levels for varying input voltages. Consider the macrocircuit, shown in Figure 11.7, of an ICL8741 operational amplifier. This was configured as a voltage gain stage (Figure 11.6), and the effect of an open-circuit at the base of Q1 was simulated using SPICE. After inspection of the circuit and further simulations at the device level it was found that three other hard faults produced identical fault effects of Q1 base open-circuit: these were Q1 emitter, Q3 emitter or Q3 base open-circuit, which immediately collapses four faults into one.

The voltage input/output characteristics of this fault is shown in Figure 11.8, and since this consists of only two linear operating regions, it is very simple to

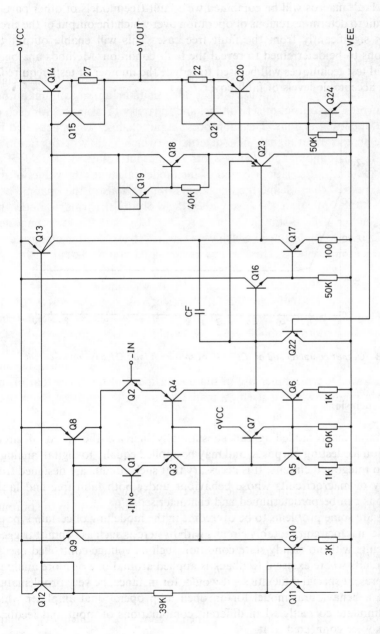

Fig. 11.7 *Circuit diagram of ICL8741 opamp.*

macromodel in exactly the same fashion as the fault-free model demonstrated in the previous section.

Such fault macros will be combined with fault-free models of other parts of the circuit to determine regions of operation over which the output of the circuit deviates significantly from the fault free case. This will enable output test conditions to be determined to reveal the fault condition. Methods analogous to digital test techniques will be used to reduce the number of tests required to achieve acceptable levels of fault cover.

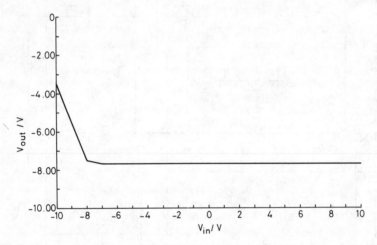

Fig. 11.8   *Output characteristic of ICL8741 opamp with fault Q1 base open-circuit.*

11.7 Conclusions

Behavioural macromodelling can substantially increase the speed of circuit simulation for testing purposes, and may be applied equally to digital, analogue or mixed integrated circuits. It is necessary that such circuits are designed from a library of macrocircuits whose behaviour under both fault-free and faulty conditions can be predetermined and characterised.

There are some problems to be overcome in the modelling of certain types of analogue macrocircuit: notably circuits with hysteresis such as Schmitt triggers, and circuits with no steady-state condition such as voltage-controlled oscillators. Circuits where external feedback is applied around one or more analogue blocks present special difficulties – it would, for instance, be very problematical to write a behavioural model for an open loop operational amplifier which would simulate correctly when different combinations of input and feedback resistors were connected to it.

The modelling of macrocircuits with hard faults is usually no more demanding than modelling the fault-free macrocircuits, as an open or short circuit often leads to a catastrophic failure at the output, which is both simple to model and fast to simulate.

This greatly eases the task of formulating test patterns for a given device as a test engineer can compile a comprehensive fault dictionary through rapid fault simulation runs which would otherwise require considerably greater computing power and time, and hence cost.

References

BOYLE, G. R., COHN, B. M., PEDERSON, D. O., and SOLOMON, J. E. (1974): 'Macromodelling of Integrated Circuit Operational Amplifiers', *IEEE J. Solid-State Circuits,* SC-9, (No. 6), Dec 1974, pp. 353–63

DOREY, A. P., SILVESTER, P. J., BALL, R. J. (1987): 'The Testing of Integrated Circuits Incorporating Analogue Sections' *in* MILLER, D. M., *Development in Integrated Circuit Testing,* (Academic Press, NY), pp. 390–405

HSIEH, H. Y. and RABBAT, N. B. (1978): 'Macromodelling and Macrosimulation Techniques,' *Proc. 1978 IEEE Int. Symp. on Circuits and Systems* pp. 336–9

HUGHES, J. L. A., McCLUSKEY, E. J. (1986): 'Multiple Stuck-at Fault Coverage of Single Stuck-at Fault Test Sets' *Proc. 1986 Int. Test Conf.,* 8–11 Sept. 1986, pp. 368–74

JHA, N. K., (1986): 'Detecting Multiple Faults in CMOS Circuits,' *Proc. 1986 Int. Test Conf.* 8–11 Sept. 1986, pp. 514–9

SILVAR-LISCO (1986): *HELIX Command Reference Manual,* Vols 1 and 2, November 1986

WILKINS, B. R. (1986): '*Testing Digital Circuits*' (Van Nostrand Reinhold, NY)

YIN, P. M. and ELCHERIF, Y. S. (1985): 'Analogue Circuits Fault Dictionary – New Approaches and Implementation,' *Int. J. Circuit Theory and Applications,* 13, pp. 149–72

DFT and ATE systems in practice

Chapter 12

Test economics in a VLSI environment

I. D. Dear and A. P. Ambler

12.1 Introduction

It is well-known that the cost of testing ICs has rapidly escalated as more circuits with increased complexity are integrated onto a single substrate (Goel, 1980). This is as the ratio of transistors to pins has increased by orders of magnitude over the last few years. As designs move onto ULSI and WSI circuits then the test access problem will carry on increasing, unless design and test styles change. There is much scope for modifying the design of complex ICs to increase test access and reduce test costs (Konemann, *et al.*, 1979, Eichelberger, *et al.*, 1977). However, designers are often very reluctant to give up too much silicon real estate for these test features, when there is so much demand for that space for functional circuits, which the designer can see as a real improvement to the product, thus increasing the products marketability.

Arriving at the best compromise, trade-offs between manufacture and test factors – having appropriate regard for the customers test needs and deciding on other relevant marketing factors – is becoming a major part of IC design activities. The main aim of this chapter is to examine test factors and how they are related to other design and manufacturing features from an economic point of view. Example cost predictions are given based on a cost model and cost related data obtained from industrial companies (Varma, *et al.*, 1984, Ambler, *et al.*, 1986).

12.2 Design and manufacturing factors

An increasing trend in modern design environments is to have agencies which are free to use a variety of different manufacturers to produce the products they design. Such agencies can the select a manufacturer that has the appropriate processing and test facilities that are best suited to product design and marketing requirements. The final choice will have much regard for cost factors, as well as factors such as time to market. At this time, however, many manufacturers have their own design teams and product designs need to be constrained to the companies' present manufacturing and test capabilities. This is unless additional production plants can be funded or the appropriately equipped sub-suppliers found. There are many supply situations and often designers need to have regard for several supply possibilities, when deciding on the best overall arrangement to manufacture and test their products.

There is often the need to consider the effect new products will have on those already in manufacture, as processing and test facilities need to be shared with other products. A big factor is to provide for the appropriate recovery of investment into processing plant and test facilities (today automatic test equipment (ATE) for ICs and systems can cost over 1 million dollars). Because of expensive processing and test facilities, it can be difficult to contract out to specialists or off-load over-committed 'in-house' facilities. The problem is the difficulty of harmonising the sub-suppliers' and main suppliers' test and other facilities. For example, labour-intensive work can often be put out to third-world countries, with the main supplier testing their own ICs. It follows with all these supply options that working out the lowest test cost option can be a difficult task as can be deciding on the best trade-offs to minimise other costs and the many compromises needed between new and old products. A key management requirement when deciding on launching a new product can be to define the real risk in both technical and cost terms.

12.3 Manufacturing and test stages

The technical and cost risk may well need to be examined at all levels of production. As ICs become more complicated with increasing integration, the added value at each processing stage can rise rapidly. If the test stage allows faulty devices through, then a great deal of money can be spent processing these defective products, which are destined to be useless unless rework is possible. Then, of course, rework can be a very expensive additional cost. There is the important point that testing of devices and fault diagnostics becomes more difficult as higher levels of manufacture are reached (a 'rule-of-tens' is often quoted as the cost difference in isolating and fixing a fault between different stages of manufacture and test (Davies, 1982)). Clearly, arriving at the best stage by stage testing is very important. When it comes to testing packaged VLSI devices the cost of ten can rise dramatically. A possible supplier's aim can be to try to do as much testing as possible before packaging (wafer test), when they have greater test access, via probe pads, to the circuits on the substrate, than post-packaging testing (component or IC test), which is concentrated on detecting mainly packaging failures, such as hermetic seals and pin to die contacts. This is only one possible solution. Customers may require packaged devices to be 'fully-tested' at least for the functions they require the device to perform. This introduces the point that different customers may want to use ICs in the same way. In some cases a better arrangement can be for the customer to do their own functional testing.

12.4 Product quality and reliability

Rarely is it possible for suppliers to afford or guarantee perfection of delivered ICs; they can and often do declare a small percentage defective in each batch supplied, particularly if large batch sizes are used. The number of defective items (quality of the batch) will of course affect the batches market price. However, finding the defective items may cost the customer more than they expected if the defective items end up on his boards or even in equipment (Davis, 1982). Furthermore, when the power is switched on, fault currents induced by the defective devices can flow through connected good components causing overheating and start or accelerate material migration in these otherwise good ICs. Then when re-working the equipment to remove faulty ICs the extra handling can inflict more damage. Often a large percentage of the re-work can be to put right shortcomings in previous re-work. Whether or not the customer accepts percentage defective batches or other acceptable quality limits most customers and ultimate users of equipment will have a need to test the encapsulated ICs and product testability can be a big marketing factor. It is important that designers have appropriate information so that they can decide on the best design and test options for a new product to cover for design validation, quality evaluation of manufactured items and provide for customer and ultimate users' test needs.

12.5 Increasing product test and testability

When deciding on the best test options there are many trade-offs well worth considering; one could improve the manufacturing and test facilities to increase the processing capability. There are many other examples of this kind, and many of these can affect the products already in manufacture as well as the new ones being designed. However, this is less the case when considering the partitioning of circuits within an IC to provide for better testability and subsequently lower test costs. The extra manufacturing and design costs to provide for the partitioning can often be recovered many times over by the savings in test generation and equipment costs. There is a point at which, by providing extra partitioning circuitry on a substrate, one can help to by-pass faulty blocks when the device is in functional mode (designers can still be very reluctant to give up substrate space). With the trends towards more complicated ICs and WSI, partitioning for test, fault avoidance, etc., is becoming more frequently a design requirement. From a manufacturer's point of view, fault avoidance during processing could provide for high product yields of devices. Fault avoidance capability within an

IC also can be very useful to those customers that need to have ICs with graceful degradation; systems in satellites and submarine cables often have a need for graceful degradation of component as do many avionic, marine and land systems.

The decision to use partitioning may not always be to help test or product reliability; it may be to provide for reconfiguration of a device's circuit blocks in different ways to meet the many different functional requirements of the device's customers. Partitioning can be helpful from the manufacturer's point of view as a way of marketing several different ICs from a common design and production line. This can be a way of overcoming the 'fall-off' in production numbers as IC devices become more complicated and embrace more of the equipment's circuitry within one sealed device. In some cases customers may want to use the ICs as 'multi-rolled' devices with the partioning arrangements switchable as part of overall system control. For this use of partitioning, test costs can rise at a very dramatic rate unless great care is taken to harmonise functional partitioning needs with those for test and fault avoidance to provide for graceful degradation of devices.

12.6 Managerial and marketing factors

With present-day electronic products, competition is rising sharply, and IC suppliers have to look closely at how to use their increasingly expensive manufacturing and test facilities; with test costs now being a large proportion of total costs more automation of test facilities is favoured by many suppliers. Recovering 0.5 to 1 million dollars for an automatic test station is not an easy matter. Also, it can be necessary to constrain designers of new products so as to make them testable by the existing companies ATE. This is partly the reason why economic families of products which need much the same production process and testing are produced. As a product in a family starts to lose its market a new product is needed to replace it. As a temporary measure the price of a product can be lowered by using marginal costing. This is to excuse a product from contributing to some overhead cost that is shared by others. This is easy to do if the product has more than paid its share of these overhead costs in a long successful production run. Also, it is important to hold a market position until a new one is available. A different consideration is to adjust the marketing of a product within a family to maximise overall profits when opportunities present themselves. In all there are many manufacturing and test factors of this kind that need to be considered when designing a new product. This is not to overlook the trend in market demand for more powerful higher technology devices (such as WSI devices), and product families will need to be appropriately advanced.

If possible, it is a great help if information can be gathered at the right level of detail to serve day-to-day management needs and to decide on strategic planning options; planning must take into account the use of expensive processing plant, achievable processing yields, and other factors. In the following examination of test costs, the information used is based upon the stage-by-stage processing of ICs from wafer products to their final encapsulation into systems and in-field use. As IC suppliers and equipment manufacturers differ greatly in the design and manufacturing facilities, mix of products, scale of production, market position, then clearly their stage-to-stage processing and test requirements/costs can differ greatly. What is presented here cannot be taken as typical but only an illustration of test cost planning that has regard for some of the wider economic factors. The main aim is to show how design for test (DFT) options of different kinds (e.g. Scan and BIST) can affect test costs. Also, how test options can appear different when viewed from different customer, supplier, or ultimate users' standpoints. This type of strategic test cost planning is very much the centre of economic test management for ICs in production today and coming on stream in the future.

12.7 Generalised test economic effects

As mentioned above there are many cost areas to consider in electronic system production today. The following are well recognised standard cost areas that need to be considered in VLSI cost evaluation:

1. Design costs
2. Processing and plant costs
3. Packaging cost
4. Manufacturing test costs: wafer, component, board and systems.
5. Field test costs.

This section is aimed at reinforcing some of those points already made in this chapter by showing the effect test options can have on the overall production and test cost for a system. As mentioned above, many factors change between design, manufacturer and company's market position. The following can thus only give an indication of the type of cost areas that might be worth considering.

Figure 12.1 shows a predicted cost breakdown for a system containing about twenty VLSI devices mounted on a few boards, for a production run of 100 systems. It can be observed that cost for packaging and wafer test can be ignored because of the low production volume. A large fraction of the system costs occurs due to initial volume-independent costs such as design costs, test pattern generation, fault simulation, processing (i.e., mask production) and plant set-up or learning time costs. The wafer test cost is dependent on whether there is any amount of functional testing in addition to basic processing. It is possible for the wafer test cost to be comparable in size to that for component testing when full functional testing at the wafer stage is performed.

Figure 12.2 indicates how the cost partitioning pattern changes for a system as the production run of the same product is stepped up to 10 000 systems. Now, such costs that are volume-dependent become more observable – i.e., packaging of components and systems, and field test costs which are based on the number of field failures. Also it becomes obvious that test costs have not reduced proportionally, relative to the volume-dependent test factors. This might be

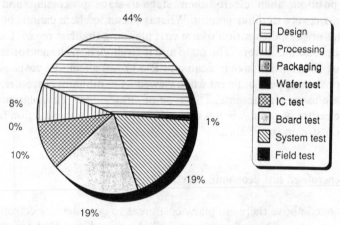

Fig. 12.1 *Example cost partitions (NO DFT).*

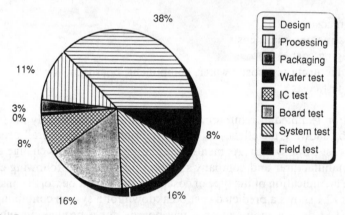

Fig. 12.2 *Example cost partitions (NO DFT).*

expected because test pattern generation (TPG), fault simulation and test program production costs that are normally considered to be the major test cost factors are amortised with test volume. This is explained by the importance of the test throughput cost factors; these factors increase because of the requirement for increased use of the expensive capital plant related to test, such as ATE, and also the increased amount of re-work or test monitoring required.

These figures show the relative importance of each standard cost area related to the stage of production and test. The cost partitioning will, however, change from product to product, between manufacturers and within the intended market. For example, it is often possible to put out to contractors much of the silicon processing and/or testing. This could reduce the expensive plant cost factor if only small production quantities are required.

The intended ultimate users can greatly influence the cost breakdown. If the system is to be used where field failures are very expensive (such as in avionic, space, and military systems) then the cost partitioning can be strongly influenced. Figure 12.3 shows how the cost partition can change as a result of increased down-time costs from $100 to $750 per hour. The increased down-time costs can arise due to system back-up replacement requirements, loss of income due to non-working or degraded systems, and many other such factors affected by a system's reliability.

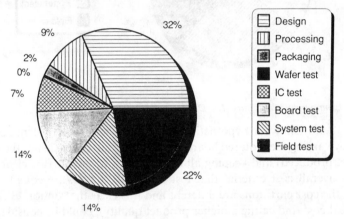

Fig. 12.3 *Example cost partitions (NO DFT).*

The increased importance of the field test to the overall system life costs, due to high down-time costs, might cause concern. Thus special attention could be paid to reducing this cost factor. The field test cost is mostly affected by the number of field failures, which is a direct result of the system's reliability. The increase of system reliability can be controlled in the following three ways: (*a*) improving the design style to incorporate redundancy and/or test aids, (*b*) having better control of manufacture processes to improve quality, and (*c*) improving the quality of the test procedure. However, each one of these will cause an increase in one of the standard cost areas described. The effect of improving test quality on the cost partitioning can be observed in Figure 12.4 where there has been increased system testing. The increased quality can be obtained by spending more on TPG to obtain higher fault coverage at each stage.

Up to now this section has concentrated on the effect that the standard cost areas (design, test, manufacturing and marketing) factors can have on relative cost partitions. However, the overall cost of a product is often the bottom line. Figure 12.5(*a*) shows how the total production and test cost changes as the level of board and system fault coverage is increased. It is clear that the product's production volume can be a key factor when deciding on the amount of test required.

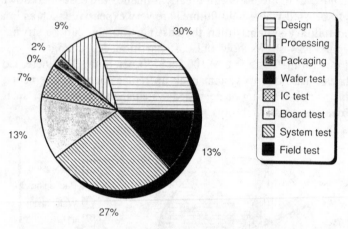

Fig. 12.4 *Example cost partitions (NO DFT).*

Figure 12.5(*b*) shows a hypothetical situation where total system costs are changed by varying the required board and system fault cover (hence the quality of product produced) and keeping all other factors constant. It is clear, based on purely overall cost criteria, that test option 'B' or 'C' are equally viable. However, the cost partitions are different and it is clear that option 'B', having a lower field cost (indicating a higher product quality), could be considered the better option. Continuing the quality and reliability point further, a choice between the 'B' and 'D' could be made; because the survival of a company depends as much on gaining market share by competitive market prices as on keeping the present market share by having a good reputation for high-quality products. It can thus be profitable in the long-term to take a cost penalty on one product and maintain a higher quality of product. Such test options require experienced marketing decisions.

There are a multitude of test and other variations possible when evaluating a product's cost options. Other factors that can have a large effect on product costing include the product mix on a process line, family grouping of products, use of common test equipment between manufacturing stages, and the use of subcontractors; it is important to consider all these factors when costing for a product but these are always highly product- and manufacturer-oriented. The factors we have looked at have been chosen to stimulate an awareness of the interaction of VLSI cost areas and not to stress the importance of any one particular area.

12.8 Economics of built-in Test

The cost saving associated with the use of BIT manifests itself throughout the complete life-cycle of the device; it much reduces or in some cases eliminates the task of test vector generation (Konemann 1979) and fault simulation, while reducing the costs of fabrication testing, board and system test and the in-service testing by reducing test times and ATE costs. It is accepted that these factors must be weighed against the disadvantage of reduced yield and higher design and fabrication costs through the use of increased silicon real-estate.

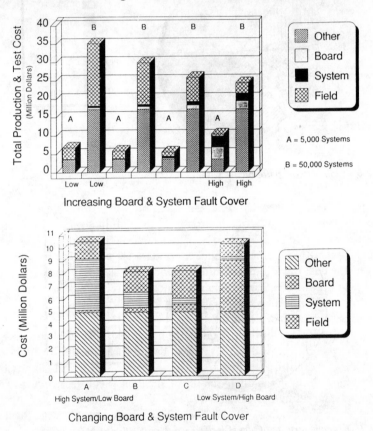

Fig. 12.5 *(a) How much test? (b) Test cost/quality.*

Figures 12.6 and 12.7 show the cost partitioning for the same system as in Figure 12.3 but employing DFT (Eichelberger *et al.*, 1979), and built-in self-test (BIST) (Konemann *et al.*, 1979) strategies respectively. Again, it must be stressed that the partitioning will also be dependent on product, manufacture, and market requirements, as well as on test stragegies. The test cost importance has been much reduced due to the reasons mentioned above. The increased silicon usage required to implement these strategies has also affected the processing cost as well as the extra design effort needed. In these cost partitions it is assumed that the DFT strategies are incorporated in each stage of system manufacture from wafer to system production and can be used when in-field testing is performed.

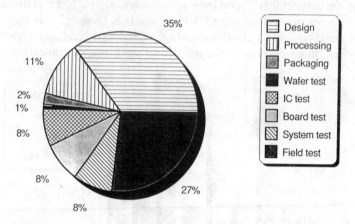

Fig. 12.6 *Example cost partitions (LSSD).*

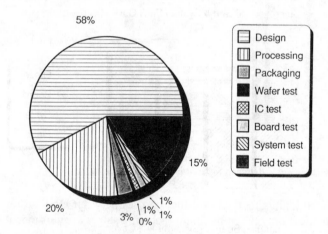

Fig. 12.7 *Example cost partitions (BILBO).*

Figure 12.8 indicates how the relative cost areas between production and test stages are affected by the test strategy adopted. The figure also indicates the effect of processing factors such as process yield can have on test options, as well as indicating the savings due to DFT. These are most noticeable at the higher levels of test and system life.

Figure 12.9 shows how the total production and test cost varies with the size of the production run, for three different test options: no DFT, LSSD, and

BILBO. The sensitivity bars show how the cost curves can change as various DFT related parameters can be realistically adjusted between design, manufacture and marketing options mentioned in the previous section.

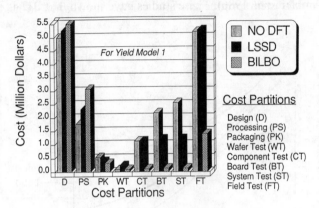

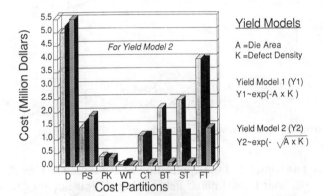

Fig. 12.8 *Relative costs between test options.*

Figure 12.10 highlights the point that choices between DFT options can not always be made if only the initial manufacturing stages are considered (i.e., that of a purely IC manufacturer). A supplier can need different test and other facilities to those of his customer. Often, therefore, the views to whether a given chip design has BIST or some other testing aids or not can vary between the IC supplier and his customer. An IC manufacturer might come to the conclusion that certain test aids are not directly economically beneficial for particular designs or production volumes. However, devices that include testing aids accessible to board and system testing can be of much cost benefit to system manufacturers. IC manufacturers could use this selling point to increase the price of their devices or increase their market share.

Much effort has been made to show how cost evaluations vary between manufacturers and designs. However, studies have been performed using specific *real* manufacturing and design data to verify the predictions and trends observed (Ambler *et al.* 1986). These studies have shown that BIT is commercially viable.

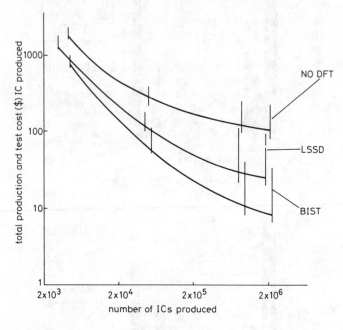

Fig. 12.9 *Cost variations between DFT options.*

The above has considered the application of the same DFT method applied to the whole IC, board, system and field testing. This is not always possible or the most cost-effective method. An option can be to apply DFT to only lower levels of test (i.e., wafer and IC) on the basis that the silicon cost can be waived against the very high assurance of product quality, thus reducing the level of test required at later more expensive test and manufacturing stages. Furthermore, the use of one specific test method is not always the best method for individual ICs; the use of different test strategies for different ICs producing a hybrid (multi-method) test strategy for the board can mean an economically viable method. This argument can be taken further to hybrid testing of ICs since no one DFT method is most suited to the testing of IC building blocks. For example, the application of LSSD to a random access memory does not always produce a very efficient test strategy because of the large set of test vectors required to be serially scanned into a device to obtain a high fault cover.

12.9   Conclusions

Companies differ greatly in the variety of products they produce, as to their economic scale of manufacture, investment into capital plant and a host of other key factors. Each company will, therefore, need to adjust the information presented in this chapter and relate their own 'in-house' analysis to the broader canvassers. This enables a company to obtain the best trend information on which to base future design and manufacture strategies.

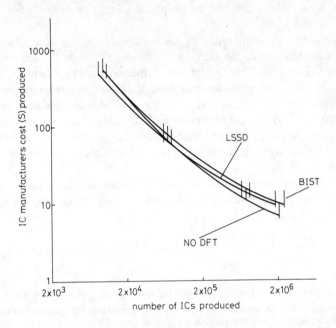

Fig. 12.10   *Cost variations between DFT options.*

For one to achieve economic yields of devices as they become more complex with large scales of integration, the ability to reduce the inherent number of faults is critical. As the number of faults per IC or wafer is reduced it becomes more economical to use built-in testing aids such as scan path or employ built-in self-test circuitry. Often the more complex and the larger the scale of integration, the fewer devices of each type are required by customers. Built-in testing aids become more attractive for smaller numbers of devices. Also, if larger quantities of devices are produced, and very low fault densities of complex devices are to be achieved, then scan path and BIST is attractive. Thus, there are several manufacturing conditions when there is an economic advantage in using BIST. However, using built-in or external testing is rarely a clear-cut decision for an IC suplier as some products will be better without built-in test and other products will be unattractive to a customer without it. Having invested heavily in external test equipment, a supplier may find it best to use external testing whenever possible. Then there is the concern that a new company can enter the market place supplying a competing range of devices with built-in testing aids that are more attractive to the customer.

For those purchasing devices to include in equipments and those that service and maintain the equipment in the field, the cost of external test can be a very large proportion of overall costs. Analysis shows that such customers can often greatly reduce their testing costs if devices have built-in test aids. In many cases built-in test may not be attractive to a supplier but much needed by the customer. Customers may well be prepared to pay extra costs for including testing aids on a chip.

Unfortunately, the difficulties of testing devices to an acceptable level of confidence are still not appreciated to a sufficient extent by designers of circuits and systems. Too often they regard it as 'someone else's problem' or, at best, as an afterthought. Such attitudes will bedevil those companies retaining these practices; the best person to make a device testable is the designer him/herself – any testability features are best considered at design time and not later. Until designers can be educated, and management re-educated, to the belief that design for testability is at least as important as functional design, then this problem will remain.

It is hoped that this chapter will at least start to cause those who have dismissed or not considered design for test and its importance in real terms (money) to wake up!

References

AMBLER, A. P., PARASKEVA, M., BURROWS, D. F., KNIGHT, W. L., DEAR, I. D. (1986): 'Economically Viable Automatic Insertion of Self Test Features for Custom VLSI,' *IEEE Int. Test Conf., Washington, USA,* Sept 1986, pp. 232–43

DAVIS, B., (1982): '*The Economics of Automatic Test*,' (McGraw-Hill, London)

EICHELBERGER, E. B. and WILLIAMS, T. W. (1977): 'A Logic Design Structure for LSI Testability,' *IEEE Design Automation Conf.,* pp. 206–11

GOEL, P. (1988): 'Test Generation Cost Analysis and Predictions,' *IEEE Design Automation Conf.,* pp. 77–84

KONEMANN, B., MUCHA, J., ZWIEHOFF, G. (1979): 'Built-in Logic Block Observation Techniques,' *IEEE Int. Test Conf.,* pp. 315–19

VARMA, P., AMBLER, A. P. and BAKER, K. (1984): 'Analysis of the Economic of Self Test,' *IEEE Int. Test Conf., Philadelphia, USA,* Nov. 1984, pp. 20–30

Chapter 13

Evaluation of design for testability methods

A. H. Boyce

13.1 Introduction

With the emergence of very high speed integrated circuit (VHSIC) technology, it is important that testability is included in the design phase. Such a practice will usually incur a penalty in terms of increased chip area usage, degradation in performance and extra pin count. However, this must be weighed against the associated costs of producing a device whose quality cannot be guaranteed. All aspects of the test process are inherently expensive, but the consequences of inadequate testing within the manufacturing process could prove to be even more expensive in the long run.

It is assumed here that the testing is being performed to confirm the correctness of fabrication and not to validate the design.

13.2 Fault models

When structural testing is to be undertaken the majority of users consider the classical 'node stuck-at' or, 'pin stuck-at' fault models on the gate circuit. Additional faults can be considered such as, shorts between adjacent nodes, stuck-on and stuck-off faults in MOS circuits. However, the number of faults that can occur in practice is much higher than this list and for the majority of circuits the representative 'stuck-at' faults is normally sufficient. Note that a circuit containing n nodes has only $2n$ possible node stuck-at faults but 2^n-n-1 possible short faults. Many fault simulation programs produce a statement that $x\%$ faults are covered, but it actually means $x\%$ of considered faults are detected, not $x\%$ of all possible faults. It is normal for one to consider pin stuck-at faults on the gate circuitry but some of these can be grouped together. If with the access available, different faults cannot be identified, then they can be considered as one fault group. For instance, a NAND gate, when any input is stuck-at 0 or when the output is stuck-at 1, it is impossible to tell which fault has occurred unless each individual gate node is monitored.

There are classes of faults that can turn a combinational circuit into a sequential function. One particular set was first described by Wadsack (1978) and occurs when the switched level MOS model is used. Difficulties now arise in that testing may require two adjacent test patterns (i.e., two patterns with only one input changed). To detect faults in sequential circuits which contain one level of sequentiality – i.e., one memory between input and output pins – may need one pattern to initialise the circuit and another to drive the faulty state to the output.

A further point to be mentioned is that if an algorithm for test pattern generation is being developed, then the order in which each fault is considered may affect the efficiency of the final testing. This is mentioned in Wilkins (1986) but quickly dismissed. The penalty of doing fault ordering is the extra time necessary to implement it but test pattern generation has to be done only once. If the total number of test patterns is reduced by ten percent, then each tested chip will cost less.

13.3 Types of testing

For VLSI it is normally considered that go/no-go testing is developed. The reason for this is that a chip once made cannot be repaired; therefore, if a fault has been detected, this is sufficient to know that the device does not work. When the go/no-go test sequence is generated by the consideration of one fault group at a time, some diagnostic information is available and this should be preserved. Diagnostic testing is much more applicable to board and system testing. To produce an efficient method multi-flow testing should be used. Here, the next test is dependent on the results of the current test. Also, the results of a test pattern are not just pass or fail but all output states noted. The more outputs available the greater choice of possible output patterns giving better diagnostic resolution.

It is also necessary to remember what one is trying to test. Sometimes it is the function of a device, which may not be fully defined, and sometimes it is the structure of the device with no regard to its function. Ideally, it should be a mixture of both function and structure if all this information is available.

A related problem of testing is whether it is done statically or dynamically. To test functionally one normally needs to test dynamically but often the limitations of the Automatic Test Equipment (ATE) will not allow a device to be tested at full speed. The majority of computer generated test patterns use the structure of the device under static conditions so, certain faults in a static redundant circuit cannot be detected without extra observation being provided. The aim of structural testing is to activate all nodes and to know that if a node does not respond correctly the effect is observable from one of the monitored points.

13.4 Test pattern generation

Whichever structure is used for a design at some stage it is necessary to actually produce test patterns and their corresponding responses. There are several methods available and these will be discussed in turn.

13.4.1 Exhaustive

Exhaustive test patterns imply that all possible combinations of input states are used. Unfortunately, even with combinational circuits this may not be sufficient as Wadsack type faults may be present. Therefore to detect all possible faults in a combinational circuit it is necessary to perform all possible changes from all possible states. A circuit with n inputs has 2^n possible input patterns and is called exhaustive testing if all 2^n states are used. But to detect all possible Wadsack type faults n times 2^n patterns may be necessary. This is n times more test patterns than exhaustive testing! Because of the ease of producing psuedo-random numbers on chip exhaustive testing is often used for circuit partitions of up to about 20 inputs. Over 26 inputs the testing time can be measured in minutes.

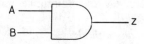

Inputs		Z	Faults				
A	B		A s.a.0	A s.a.1	B s.a.1	Z s.a.o	Z s.a.1
0	0	0	0	0	0	0	1
0	1	0	0	1	0	0	1
1	0	0	0	0	1	0	1
1	1	1	0	1	1	0	1

Fig. 13.1 *Two-input AND fault matrix.*

13.4.2 Functional

For simple functions, particularly for combinational circuits, functional testing can be useful. This can be illustrated by taking a two-input AND gate. Figure 13.1 shows the fault matrix of a two-input AND with all pin stuck-at faults considered. B stuck-at 0 and Z stuck-at 0 are collapsed into fault A stuck-at 0. From the fault matrix fault A stuck-at 1 can only be detected by input $AB/01$ and fault B stuck-at 1 can only be detected by $AB/10$. Similarly, fault Z stuck-at 0 is only detected by $AB/11$. Fault Z stuck-at 1 can be detected by any test pattern except $AB/11$. Therefore to test this two-input AND gate the patterns 01, 11, 10 would be sufficient for the faults considered. An AND gate with n

inputs needs $n + 1$ test patterns not 2^n. However, if n is sufficiently small 2^n test patterns would give greater faults coverage than $n + 1$ and still be practical.

For complex functions the generation of test patterns is not so easy, and it is necessary for the designer to specify the behaviour of the device. It is interesting to note that if a complex function is given to x different designers there will be x different designs. The functional tests should be equally useful to all x designs. Often it is found that a design does more than the function required of it and that more efficient test patterns can be generated using these additional functions. A particular case in point concerned a three-team quiz game circuit where the first person who knew the answer to a question pressed his button. Once pressed, any combination of buttons pressed was inhibited until the quiz master reset the system. A lamp indicated which button was pressed first.

The functional test patterns can readily be defined where each button is pressed first and the others pressed in a specified sequence. The design, however, performed another function which was that if all team members pressed their buttons and the quiz master reset the system then the last person to release the button had his light switched on. In this case it provided a much better test sequence. As a general rule testing is much more efficient when maximum activity occurs in the circuit with minimum number of input changes.

13.4.3 Random and ad hoc methods
The use of random generated patterns should not be mixed up with exhaustive testing. This often happens because of the method of generating the patterns. Exhaustive testing often use a pseudo-random number generator through its complete cycle, whereas random testing only uses a limited set of possible patterns. The method is not suitable for circuits with high fan-in AND gates and is only used as a last resort. Ad hoc methods are very similar to random methods but include the ability to modify the circuit without using any real test strategy.

13.4.4 Deterministic
The most used methods are based on work of Roth (1966). This is known as the D-Algorithm and requires the structure of the circuit and the fault groups that need to be considered. A fault group is selected and then the input patterns are decided to generate the opposite state on the faulty node and then to pass this fault to an observable node. D is used to denote that a node is in the 1 state when the circuit is working correctly and is 0 when the selected fault is present. The simple circuit shown in Figure 13.2 will be used to illustrate the D-Algorithm. To find a test pattern to detect input A stuck-at 0 it is necessary to put A to 1 (backward simulation). Therefore node A is defined as D (i.e. working correctly it should be 1 but with the fault present it is 0). To propagate this D through the first NAND gate it will be necessary to have the other input to that gate as 1. Since if it was 0 then the ouput of the gate is 1 and D (or $\bar{D}$) will not get to an observable node. To make this other input 1, C has to be 0. The output of NAND gate 1 is $\bar{D}$, and it must now be propogated through NAND gate 3,

which implies that its other input must be 1. However, since C is 0 gate 2 output is 1, the output of gate 3 is D. This gives the situation that the fault input A stuck-at 0 is detected by the input pattern $ABC/1XO$ and the output should be 1 and if the fault is present the output is 0.

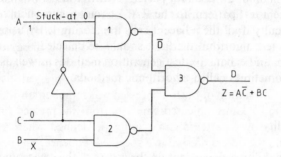

Fig. 13.2 *Simple circuit to illustrate the D-Algorithm.*

This example illustrates several points. The input B does not affect the test pattern, so B can be either 0 or 1. This fact should be remembered, as B could be specified for another fault and these two test patterns could be coalesced. If gate 2 was an AND gates instead of a NAND, then the condition input C is 0 makes the wrong conditions on gate 3 and $\bar{D}$ cannot be propagated to the output. It is then necessary to investigate any other possible test routes. However, in this example with gate 2 an AND gate the circuit contains a static redundancy as the output is then the NAND of inputs B and C and input A is not required, so any faults on A cannot be detected. If a particular fault cannot provide a solution for a test pattern then the circuit is statically redundant. The final point is when a fault is considered on a fan-out node. To pass a $D(\bar{D})$ through an AND gate (or a NAND gate) the other input has to be either a 1 or a $D(\bar{D})$. Similarly, to pass a $D(\bar{D})$ through an OR gate (or a NOR gate) the other input has to be either a 0 or a $D(\bar{D})$.

An alternative method to generate test patterns on combinational circuits is known as Boolean differences, (Amer and Condulmari 1967). The circuit shown in Figure 13.2 will be used again but with the fault C stuck-at 0. The Boolean expression for the working circuit is $Z = A\bar{C} + BC$ and when the fault is present the expression reduces to $Z = A$. If these two expressions are exclusively ORed one obtains $\bar{A}BC + A\bar{B}C$ which gives two test patterns 011 or 101. Either of these test patterns will detect the fault. By considering another fault, that is the connection of the inverter to input C stuck-at 0, the output expression is given as $Z = A + BC$.

$(A + BC) \oplus A\bar{C} + BC$ which reduces $A\bar{B}C$ (which is one of the two patterns from the previous fault).

Goel and Rosales (1981) have developed extensions to the above methods by an algorithm named PODEM.

13.4.5 Algorithmic
Another method of test pattern generation is to use a simple algorithmic technique without considering the circuit. The most obvious use is for random access memory (RAM) testing where walking 0 and walking 1 tests are performed. Such methods are used for pattern-sensitive faults when it is possible to readily produce a test pattern. In the RAM circuit a particular cell storing a 0 may have difficulty if all the adjacent cells are storing 1. By using a walking 0 and walking 1 test algorithm the fault should be detected.

Both random and exhaustive test generation methods as well as deterministic methods are sometimes called algorithmic methods.

13.5 Accessibility

The ease of testability is dependent on the access of the device under test. There are two parts to the access and both are required. Firstly,one must be able to control the inputs and to initialise the sequential circuits (i.e., control the memory) and secondly, to monitor the nodes where the faults can be seen (i.e., observability). To this end many of the successful circuit structures that can be tested easily use the principle of 'divide and conquer'. By splitting the circuit into manageable partitions and allowing sufficient access to these partitions enhances the chances of testability and allows the use of test pattern generation methods described in §13.4. It is often stated that the difficulty of test pattern generation increases as the cube of complexity. Hence, by dividing a circuit into ten equal size partitions this can reduce test pattern generation costs by a factor of one hundred times.

13.6 Design for testability (DFT)

The idea of design for testability is that testing is considered from the outset of design. This makes the designer realise that test is an important aspect of design instead of leaving all test decisions to a test engineer. With a little bit of forethought testing can be easily achieved as the designer has to consider test strategy throughout the design phase. When the appropriate test strategies are used for a design the cost of developing test patterns can be kept to a minimum. This in turn will lead to a product of high quality with known quality assurance without resorting to a percentage fault coverage figure to an unknown base.

Testing can be considered in two parts: first, get the structure right to provide the access; and second, generate the test patterns for the various partitions. It must be realised that DFT will incur additional costs, there is a silicon overhead and a degradation of performance due to increased propagation delay times.

13.7 Structural designs

13.7.1 Scan path

The most successful structured test strategy is called *scan path*. The idea is simple and the majority of designs can be adapted to this method. In scan path testing all memory latches are connected together to form one serial-in, serial-out shift register. This is achieved by adding to each latch a two-to-one multiplexer to the data input pin so that the controls on the multiplexers select either the Q output of the previous latch or the normal designed latch input. The circuitry driving the normal latch input must be combinational with primary inputs or Q and $\bar{Q}$ outputs of other latches in the scan-chain. The generalised structure is given in Figure 13.3. Three separate access nodes are required, scan-in, scan-out and test/normal. Both scan-in and scan-out could be multiplexed with other output and input pins, respectively, but the test/normal is an essential extra pin as this has to inform the chip that a test mode is to be instigated.

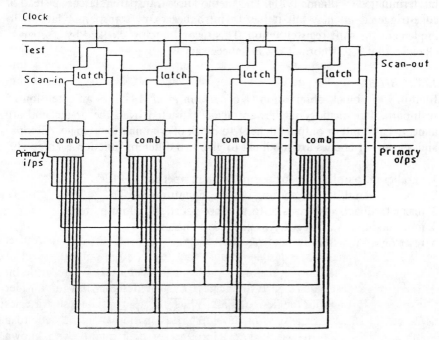

Fig. 13.3 *General structure for a scan path design.*

The structure naturally divides the sequential parts from the combinational parts. The individual combinational tree that drives their respective latches can also readily be determined. It is necessary to produce the test-pattern of each individual combinational tree using a previously described method, e.g. D-algorithm. These test patterns are then merged so that as many trees are tested in parallel as possible.

Testing is performed by initialising the circuit and checking that the serial-in serial-out shift register is operating correctly. (Note that if this fails the device is rejected even though functionally the device may work correctly. This is because an extra function has been added to the device for testing purposes and if this does not work the device cannot be tested.) The test patterns are then entered serially via the scan-in pin and at the appropriate time, the test/normal pin is changed to select the output from the appropriate combinational trees and a further clock pulse given. The results can then be shifted out via the scan-out pin and at the same time the next test pattern shifted in.

An adaptation of scan path developed by IBM is known as *level-sensitive* scan design. Logic is level-sensitive if the steady state response to any allowed input state change is independent of the circuit and wire delays within the logic. Also, if an input state change involves the changing of more than one input signal, then the response must be independent of the order in which they change. Steady state response is the final value of all logic gate outputs after all change activity has terminated (Williams 1981). This method uses transparent latches instead of edge-triggered latches, but it has to introduce extra transparent latches to implement the shift register action. This second latch is clocked by a separate clock so that a test/normal pin is unnecessary.

13.7.2 BILBO

Built-in logic block observation (Koenemann *et al.* 1979), is an extension of scan-path. The memory latches are again extracted from the design and are allocated as input or output latches to a block of combinational logic. All input and all output latches are modified to give five different functions. There are:

1 parallel-in, parallel-out for normal system operation (PIPO)
2 serial-in, serial-out for shifting out the signature (SISO)
3 linear feedback shift register to produce psuedo-random patterns
4 multiple-input shift register to produce signature
5 reset on all registers.

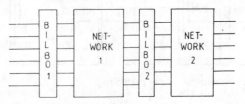

Fig. 13.4 *General structure for a BILBO design.*

The structure of a BILBO design is shown in Figure 13.4. To test combinational circuit 1, BILBO 1 and 2 are reset and BILBO 1 seeded using the SISO function and then put into the pseudo-random generator mode to feed data into the combinational network 1. The results of these patterns are fed to BILBO 2 while it is in the multiple input shift register mode. When the necessary test patterns have been applied – i.e. after a specified number of system clock pulses – BILBO 2 is put into SISO mode and the signature shifted out. These results are compared with the results obtained from computer simulation or from a known good device. Combinational network 2 can now be tested by reversing the roles of BILBO 1 and BILBO 2.

13.7.3 Random access scan
An alternative test method to scan-path is random access. Instead of connecting all memory latches into a serial-in, serial-out shift register, each memory is separately addressable using the address X and Y. Figure 13.5 shows the circuitry for each latch so that the user can address any latch and write into or read out of each latch. Shadow latches have been added so that a sequence of test patterns can be applied that are essential to test sequential circuits. Test data and test data mon. correspond to scan-in and scan-out, and test/normal provides the same function as in scan path by selecting either the normal data input or the forced value previously entered via test data on to the latch.

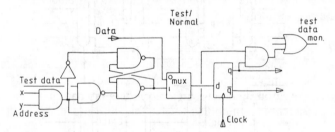

Fig. 13.5 *General structure of a memory for random access.*

The overhead looks high because of the additional shadow register for each latch but this would also apply to scan path if it was necessary to have a fixed sequence of test patterns without extraneous patterns between them. To set a latch to a specified state both the X and Y address lines must be 1 and the value entered via the test data input. This input is common to all shadow latches. The value is stored in the cross-connected NAND gates and further shadow memories can now be addressed. When all necessary shadow memories have been set to their desired states test/normal is put into test mode (i.e. 1) and a clock pulse given to transfer the stored states to all their corresponding latches.

To observe the state of any latch it is only necessary to address the latch, again using X and Y address lines and the value observed on the common test data mon. output pin. Random access has the advantage of not having to shift all input and output through a serial shift register, but when many combinational trees can be tested in parallel (i.e., with all trees are similar complexity) this advantage is very minor. The disadvantage of random access is the necessary extra circuitry needed to address each latch and the input pins this requires.

13.7.4 Boundary scan

The previously described circuit structures all deal with the problem of testing the final chip, but boundary scan is concerned more with the testing of the board. This is an important problem, and boundary scan, when employed in the chip, can make board testing easier. In essence all input and output pins of the chip have added latches, (see Figure 13.6), so that they can be connected together as a serial-in, serial-out shift register or as parallel-in, parallel-out registers for their normal working mode. Further chips on the board which contain boundary scan can now have all of their registers connected together in one big shift register. This allows, at board level, entry to any chip input and to monitor any chip output. The method also allows one to test the interconnection between chips even when other circuitry not using boundary scan is inserted between them. The method is particularly useful for built-in self-test chips. The disadvantages are that all boundary scan latches are additional and that extra shadow registers may be required if chips are not readily testable.

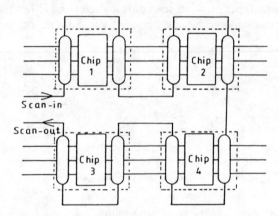

Fig. 13.6   *General structure for boundary scan.*

13.8 Conclusions

Testing must be considered early in the design phase and the test strategy selected before final circuitry is produced. The majority of successful test strategies all use the principle of 'divide and conquer'. Whenever possible analogue circuitry should be considered separately from digital circuitry. Any asynchronous design increases the testing problem, and therefore one should either avoid it or modify the circuit to allow more direct access. When scan path test strategy is chosen and the rules obeyed test pattern generation can be guaranteed. With good automatic test pattern generation methods, fault simulation is not necessary other than to discover the coverage of unconsidered faults. Fault simulation can be more expensive than test pattern generation. Finally, always attempt to provide easy circuit initialisation and that test patterns are independent of each other (i.e., no test sequences or homing sequences).

References

AMER, V. and CONDULMARI, V., (1967): 'Diagnosis of large combinational networks', *IEEE Trans. Electron Computers* **EC16**, pp. 675–80

GOEL, P., and ROSALES, B.C., (1981), PODEM-X: 'An automatic test generation system for VLSI design structures'. *Proc. 18th Design Automation Conf., June 1981*, pp. 260–8

KOENEMANN, B., MUCHA, J. and ZWIEHOFF, G., (1979): 'Built-in Logic Block Observation Techniques', *Digest of Papers, 1979 Test Conference, Oct. 1979*, 79CH1509-9C, pp. 37–41

ROTH, J. P., (1966): 'Diagnosis of automation failures: a calculus and a method', *IBM J. Res. Dev.*, **10**, pp. 278–91

WADSACK, R., (1978):'Fault modelling and simulation of CMOS and MOS IC's', *Bell Syst. Tech. Jn.*, (1986): **57**, (5), pp. 1449–74 (June)

WILKINS, B. R., (1986): '*Testing Digital Circuits – An Introduction*' (Van Nostrand Reinhold, UK)

WILLIAMS, T. W., (1981): 'Design for Testability', ' Edited by ANTOGNETTI, P. *et al.*, '*Computer Design Aids for VLSI Circuits*', *NATO Series E Applied Sciences* No. 48, pp. 359–416 (ISBN 90. 286-2701-4)

Chapter 14

Implications of ATE for ASIC Designers

C. J. Ellingham

14.1 The verification and testing of ASICs

ASIC design verification occurs in two phases. The first phase is a preliminary verification which is performed using CAE simulation tools. By making simplifying assumptions, the simulator helps the designer decide whether the proposed design has a reasonable chance of working. Based on this information, the designer decides whether it is worthwhile to make a few prototypes.

The second phase is the hardware verification process, in which the designer determines whether the prototypes, and the design in general, meet the design requirements. Since it deals with actual prototypes, instead of simplified models, it is this hardware verification which constitutes the final design verification.

Once an AISC design has been committed to manufacture the problem of production testing must be addressed. A percentage of the dice on a wafer will not function correctly because of defects which have arisen during the fabrication process. The purpose of testing is to identify such defective dice as quickly and economically as possible by means of a go/no-go test. Thus, by contrast with hardware verfication, testing is concerned with checking an individual die rather than with checking the design.

Both verification and production testing are becoming more difficult as feature sizes decrease and circuit complexity increases. In-circuit testing of ICs, by mechanical probing, is uneconomic in a production environment, and, with the decrease in feature sizes, is becoming increasingly difficult even for verification and debugging*. This means that all test access must be via external pins, and so the cost of testing VLSI chips has increased with increasing design complexity both absolutely, and as a proportion of the total cost of production. There are two reasons for this: the number of gates on a chip is increasing, and the number of pins per gate is decreasing. The consequence is an increase in the

---

* E-beam testing is an alternative approach of increasing interest for prototype verification. Here the electron beam of a scanning electron microscope is used in place of a mechanical prober. This technique can be used for small geometry devices and has a number of other important advantages (Richardson, 1987).

logic and sequential depth of chips causing the cost of testing to rise approximately exponentially (Schwaertzel, 1983).

This problem of rising test costs is particularly acute for ASICs which typically have short production runs, and therefore there are few devices to amortise the test pattern development costs across. There have, therefore, been many techniques reported in the literature to reduce such test costs (Bennetts, 1983). There is less information, though, in the literature on the implications of ATE for the ASIC designer. However, I believe that it is important for designers to appreciate these implications for two reasons. First, ATE facilities are expensive to purchase and use; improvements in an ASIC design which reduce the ATE facilities required are therefore of considerable economic importance. Secondly, for ASICs the 'time to market' is of paramount importance: ignoring, at the design stage, how a device is to be verified and tested may lead to expensive delays.

14.2 A review of ATE

There are a large number of types and models of ATE available in the market, which differ in the details of their operation but share certain common principles. There are four main categories: *verification testers, low-end general purpose testers, high-end general purpose testers*, and *ASIC production testers*. Despite the range of capabilities, and prices, that these categories offer they share many common characteristics.

All ATE machines operate under the control of a CPU – increasingly the CPUs used are standard workstations. The actual testing may be considered as being performed by two separate parts. First, a DC parametric unit (also known as a *precision measurement unit*) which can measure voltages and/or currents very accurately. This unit is used for detecting open and short circuits, leakage currents, etc. – which can be done when the IC is not in its normal operational mode – as well as for measuring parameters such as output sink and source current and tri-state leakage, etc. – which require the circuit to be operating normally. The DC parametric unit tends to be very slow (say, one measurement every millisecond), so that, ideally, it should be used as little as possible. More modern testers may offer multiple DC Parametric Units to speed testing throughput. There may be as many as one unit per pin (a characteristic of 'tester per pin' architectures), or a unit may be shared between, say, eight pins. Parametric units are standard components on production testers, but may be an optional extra for verification testers.

The other tester unit performs the functional tests, which are primarily used to check that the logic in the circuit has been fabricated correctly. Simplistically it can be considered as a set of very long shift registers, one for each pin on the IC. All the shift registers are clocked synchronously, and the data at the output of each shift register may be either driven to a circuit input or may be compared with the state of the circuit output. If a comparison is not true, then the circuit is faulty. Both the input drivers and output comparisons are implemented on what are referred to as pin electronics cards.

The above is a very simplified description of a functional test unit. In practice this is complicated by facilities for generating more complex input waveforms than the non-return-to-zero format described above (e.g., return-to-zero format), facilities for controlling when comparisions on the output pins are made, and for masking the results of output comparisons. Both the more complex input waveforms and the control of comparison periods are typically based upon 'shared timing resources' in which timing generators generate edges at a user-controllable period after the master synchronising clock. These timing edges will be available for use by the individual pin electronics cards, but there will be a limited number of such edges available.

Nevertheless, a simple picture of a functional test unit is one of a synchronous machine capable of generating a set of input waveforms and comparing the circuit outputs with the expected waveforms. In general the speed of operation is fairly fast – up to 50 MHz, for instance – so that the test can be performed in a relatively short time. The depth of the shift-register behind each pin can vary considerably: for a verification tester it may be as low as 1K, but for a high-end general purpose tester may be as high as 1M or even greater.

Information from ATE manufacturers on the architecture and performance of their machines is easily available, either directly from sales departments or in the literature (see, for example, Garcia, 1984).

14.3 Matching ASICs and ATE

In this section we will consider some of the consequences of the above ATE architecture for the ASIC designer who has to consider verification and test. Competition between manufacturers in the ATE Market frequently revolves around claims of 'faster, wider, deeper', and it is these issues which are the most obvious areas of concern for the designer and test engineer.

14.3.1 Pin count

The electronics cards required by the functional testing element of an ATE machine are expensive, so an obvious way of reducing ATE costs is to reduce the number of pins used during functional testing. Several of the structured design for test methodologies achieve this: for instance, scan design – where in test mode the storage elements on a chip are connected as a shift register – only requires two inputs, scan-in and test clock, and one output, scan-out, during scan-mode testing (Eichelberger and Williams, 1977).

However, there is still a requirement to carry out parametric testing on the remaining pins and to prove that these pins are connected to the areas tested during functional testing. One means of providing the necessary control and observation over the remaining pins for the parametric tester, without requiring all the other pins to be controlled, is the JTAG boundary scan proposal (Maunder and Beenker, 1987). In boundary scan serial scan cells are added to the chip adjacent to each functional pin. This permits signals at chip pins to be controlled and observed using scan methods for an overhead of four pins. While the JTAG proposal is primarily intended to provide a structured means of controlling the board test problem, it can also be exploited in a number of other ways. For parametric testing the functional test unit would control and observe these four pins, while the parametric unit would be applied to each of the other pins in turn.

However, such structured design-for-test methodologies do not assist verification or normal mode testing, where, naturally, access to all pins is required. If the proposed number of functional pins on an ASIC exceeds the capacity of the verification tester, this should be considered at an early stage in the design process. If it is not possible to reduce the number of pins (e.g., by replacing a parallel input with a serial input), then a two-pass test and additional test facilities on the ATE fixture will be required. (Such extra facilities on the fixture would allow pin electronic cards to be switched between different pins on the ASIC.) In the case of input pins which are not driven directly by the functional test unit during one phase of the test then facilities must also be provided to hold them at an appropriate voltage level throughout that test phase. This also demands that the logic design is partitioned in such a way to make this possible.

Such arrangements, though, suffer from two major disadvantages besides the obvious inconvenience and increased test time of a two-pass test. First, the switching arrangements will normally be relay-based, and hence there is an increased likelihood of problems during production test of equipment failure. Second, the addition of extra components, and hence of additional delays on some signal pins, will result in distorted timing relationships between the signals on different pins. For high-speed components this can be a source of considerable problems.

14.3.2 Vector depth

Designs which require long vector sequences to test them may give problems as functional sequences which exceed the memory depth on the ATE require repeated loading of the memory. This means that 'stay alive' facilities for the ASIC will be required if its current state is to be preserved while the re-load is in progress. If the necessary facilities to hold pins at set voltage levels while a memory reload is in progress are not available on the ATE, then external circuitry on the test fixture is required. As discussed in § 14.3.1, such additions to the test fixture are best avoided.

Instead it is better to consider adapting the ASIC design to reduce the depth of functional pattern required. There are two broad design approaches to achieve this aim: structured and ad hoc. In the structured approaches a consistent strategy is supported by the CAD design system to reduce both the test generation effort required and the complexity of the resulting patterns. Examples are scan-design (see above) and the various self-testing schemes based on the use of on-chip data generators and analysers. In the ad hoc approaches the designer inserts additional control or observational facilities to ease the testing problem for 'difficult' structures such as embedded memories and deeply sequential areas of the design like counters. (For a survey of both approaches see (Williams and Parker, 1983.)

Where the structured approaches are used it is important that the ATE used should have the necessary facilities. The self-test schemes typically require that the ASIC be initialised, then held in a self-test mode for a large, but fixed, number of cycles, before finally accessing the internally-generated 'signatures' of the data analysers and comparing with the predicted signature. In the self-test-mode no sequences on the input pins are required; the input levels remain unchanged while the clock signal is taken through the required number of cycles. It is therefore wasteful of ATE resources to use the pattern memory to apply the self-test mode directly when it suffices to apply the same pattern by looping on one line of pattern memory for the required number of cycles. However, if this looping facility is not available on the ATE, then, again, stay-alive facilities may be required while re-loads of memory are performed.

In scan testing, the functional test unit has to drive only two pins and observe one output pin, but the total test length can be quite lengthy. Again, therefore, it is wasteful of ATE resources to use the pattern memory to apply the scan test vectors directly as the majority of the columns in the memory will not be used while a number of memory loads may be required. Some ATE machines, however, offer a facility of 'chaining' memory columns together; here columns in memory may be successively applied to the three pins in use. Hence the memory is exploited more efficiently, and the number of reloads of pattern is reduced.

14.3.3 Speed

Certain physical defects on an ASIC may result in a gate operating in a manner which is logically correct, but slow – a 'delay fault' (Banerjee, 1983). Such delay faults may be of importance in any system expecting a signal to be present within a certain time. It is therefore important to test ASICs at the speed which they are specified to operate at. However, many of today's technologies result in devices that operate faster than any available ATE. So, testing at a slower speed may result in defective ASICs passing.

For dynamic design styles with a single clock it is possible to run a two-pass test at a reduced clock cycle time. In the first pass the clock low time is reduced to the specified minimum, while in the second pass the clock high time is similarly reduced. Since, in dynamic design styles, each transistor should only operate during one clock phase, this technique should stress each transistor. (Similar techniques can be used for more complex schemes.) However, this will not cover all possible delay faults.

For static design styles where the ASIC's specified operating speed is faster than the ATE a different approach is required. A possible technique, for simple designs where critical paths have only one latch between input and output pins, is to measure the delay paths inherent in the device. Since such a device will not operate correctly at a cycle rate that is faster than the set-up time plus the propagation time (or hold time – whichever is greater) from the clock to the appearance of the output. These parameters – set-up time, propagation time and hold time – can all be measured by running functional patterns and adjusting the placement of edges in the input signals concerned and of the sampling time on the output pin (Integrated Measurement Systems, 1986).

All the techniques described above, though, depend upon the accuracy of the ATE's timing system across both data generation and data acquisition channels at the device pins.

14.3.4 The synchronous environment

The ideal test environment for any component is one which matches the system environment in which the component will be used. However, the environment which can be created on the ATE is *not* necessarily the system environment. It is, therefore, necessary for the designer to consider carefully the consequences of the synchronous ATE architecture described above.

The most fundamental consequence of the architecture is that the ASIC must be synchronised to the ATE. So, for example, if a divider, or other clock recovery circuit, is used to generate an internal clock for the ASIC, it is essential that the circuit is initialisable so that the internal clock is fully synchronised to the ATE.

Similarly, since all comparisions on output pulses are related by user determined delays to a master clock, all output pulses should be similarly generated. Output pulses which occur after a delay which is insufficently determined by logic simulation pose problems for the test engineer: when is the comparator strobe to be active to catch a 'good' pulse? Here it must be noted that the ATE is not an oscilloscope – it cannot tell you *what* the behaviour of the output signals is, only if they *match* the expected good machine response. Hence, debugging is a slow process because determining what is happening has to be done by indirect means.

Pulses of this type generally occur in asynchronous systems on paths of considerable logical depth where the uncertainty in individual delays cumulates in a large uncertainty on the output. Logic simulation attempts to model all sources of delay on signal paths, but a purely digital model cannot always accurately model the complicated physics of the real world. Hence, the solution is to design synchronous systems, and to pipeline long data paths where this is acceptable. Frequently it is the throughput of a system that is important and not the absolute delay between an input signal and the resulting output.

Another source of difficulty on ATE is hazardous designs, which give glitches on the outputs. If a glitch occurs during an active comparator strobe, whether or not this is significant for the application, the ATE will fail the part. And again, since ATE is not an oscilloscope, finding the glitches and determining if they are significant for the application is a slow process.

Hazards of this type generally occur in asynchronous systems on unclocked feedback paths. (Again, logic simulation attempts to model possible hazards but cannot always succeed.) Hence, the ideal solution is again to design synchronous ASICs, and to ensure that all closed signal paths pass through a clocked element.

14.4 Conclusion

For ASICs and the systems which use them to reach the market on time and within budget, it is essential that the designer considers the problem of testing and verification throughout the design cycle. Besides considering the problems of test pattern generation the designer should be aware of what ATE resources are available, and should consider the implications of this ATE for his design work.

References

BANERJEE, P. (1983): 'A Model For Simulating Physical Failures in MOS VLSI Circuits', *MSc Thesis, University of Illinois*

BENNETTS, R. G. (1983): *Design of Testable Logic Circuits* (Addison-Wesley, London)

EICHELBERGER, E and WILLIAMS, T. (1977): 'A Logic Design Structure For LSI Testability', *Proc. Design Automation Conf.* pp. 462–8

GARCIA, R. (1984): 'The Fairchild Sentry 50 Tester: Establishing New ATE Performance Limits,' *IEEE Design and Test of Computers*, Vol. pp. 101–9 (May)

Integrated Measurement Systems (1986): 'Verification Solutions: A guide to ASIC Design Verification; *Integrated Measurement Systems – Part Number* 910–0061–002, pp. 47–55

MAUNDER, C. and BEENKER, F. (1987): 'Boundary Scan: A Framework for Structured Design-For-Test; *Proc. Inter. Test Conf.* pp. 714–23

RICHARDSON, N. (1987): 'E-Beam Probing for VLSI Circuit Debug', *VLSI Systems Design*, (Aug.)

SCHWAERTZEL, H. G. (1983): '*Testing of VLSI Circuits*', *VLSI '83* (North-Holland, Amsterdam), pp. 21–33

WILLIAMS and PARKER. (1983) 'Design for Testability – a Survey', *Proc. IEEE* Vol 71. Jan 1983 pp. 98–112

Fault-tolerance design for VLSI

Chapter 15

A Unified Error Detection Scheme for ASIC design

I. Sayers and G. Russell

15.1 Introduction

The problem of testing has always been a major issue in the design, fabrication and use of digital circuits. With the advent of VLSI these problems have been aggravated by several factors in addition to those introduced by the increase in circuit complexity.

1. The prolific use of VLSI circuits in applications ranging from standard consumer products to critical commercial controllers, has necessitated the development of more sophisticated and efficient CAD tools to contend with market demands. The most significant advances in CAD tool development have been in the areas of simulation and layout; regrettably improvements in the critical area of test have not been as dramatic. An unfortunate consequence of this situation is that inexperienced designers can now produce extremely complex but untestable chips.
2. Traditionally the problems of testing, particularly at board level, have always been neglected until the final stage of the design/implementation cycle. This has resulted in a view that a dichotomy exists in the design process in which design and test are considered as separate and virtually independent tasks. To a large extent this view is promoted by companies which assemble printed circuit boards from discrete components, as many of the problems of testing difficult boards have been solved by the simple expedient of the test engineer adding extra test points to the critical areas of the board; if necessary this operation could be performed in a post design phase although this practice is not to be recommended. Unfortunately these 'ad-hoc' testability enhancements cannot be transferred to the VLSI arena since adding extra test points to a fabricated circuit is impractical and subsequently testing must be considered as an integral part of the design process.

It is considered that the problems of testing VLSI circuits may, if not be totally eradicated, at least be alleviated to some extent by:

1. Educating the new designers of VLSI circuits in techniques for testing complex devices.
2. Incorporating the test strategies into the CAD package being used by the designer, thereby encapsulating the extensive knowledge of the test engineer within the software and enforcing a particular test strategy on the designer.
3. Supplying the designer with cells that have 'built-in' test structures.

It is considered that a combination of all three solutions should be utilised to overcome the *test man's burden*.

Although the test question is very serious and likely to remain so with the ever increasing complexity of successive generations of chips, there is one problem which is often overlooked – namely, the reliability of a new device and its intended application. If the device is to be used in a 'life-endangering' situation, such as in the aerospace industry, railway signalling or medical equipment, then the device must be able to tolerate fault conditions and produce a safe response. The layman's view that with an increase in the scale of integration there is a corresponding increase in the reliability of the device has been shown to be incorrect by the majority of research work (Fantini, 1984, Tasar and Tasar, 1977, Savaria *et al.*, 1984, May, 1979, Mangir and Avizienis, 1980, Liu, 1984, Tendolaker and Swann, 1982) carried out into reliability problems. The main problem appears to be the increase in temporary errors due to power supply interference, crosstalk on routing lines and charged particle strikes. The current concensus of opinion seems to indicate that unless steps are taken in all aspects of integrated circuit design to improve the situation reliability degradation will become a major obstacle in producing technologically advanced components i.e. breaking the sub-micron barrier for silicon. The authors' view of this problem is that techniques to enhance the reliability of circuits have been around almost as long as large electronic computer technology has existed. The problem appears to be that reliability enhancements are expensive to implement, not only in terms of cost but also in the extra effort that needs to be expended in design time for what is already considered as a 'mystic art' by most designers.

The preceding discussion has highlighted two major concerns of the integrated circuit design community:

1. The problem of increasing test complexity in the form of test pattern generation, test hardware and the difficulties of reaching deeply embedded sections of a circuit.
2. Reliability problems that will occur in the next generation of sub-micron VLSI processes.

Obviously attempts to solve both problems could be made in isolation. Indeed several major chip manufacturers now include testability enhancements in their proprietry cell libraries. The enhancements range from simple boundary scan latches embedded in the pad ring to complete BILBO structures included in each major cell. However, to date, very few IC manufacturers have taken up the challenge to incorporate fault tolerance directly into the chip. The chips that have been designed are mainly targeted at specific market areas, such as large computer telephone switching networks. A technique which encompasses both problems would optimise both the extra hardware and time required to formulate the design. One technique that has been extensively researched at Newcastle is the use of *residue codes*.

In order to reach a decision on which enhancement is the most suitable it is necessary to have an indication of the likely implementation costs for each of the techniques. The costs to be considered should include the following, however certain constraints might be added or amended in the light of the particular application:

1. Is extra silicon area required to enhance the design? This is a contentious constraint, since it could be argued that the 'extra area' is not 'extra' but part of the circuit and could thus form part of the device specification. This would be the case for circuits with built-in fault tolerance.
2. Will extra pins be required to access the testability enhancements?
3. What reduction in the speed of operation will be incurred due to any additional circuitry?
4. Are there constraints on the designer which will inhibit his ingenuity ?
5. Can the chosen approach be implemented in a CAD tool ?
6. Is the theory employed easy to understand ? This could have a direct effect on the practical aspects of the implementation.

This chapter discusses the use of residue codes to both enhance the testability of a circuit and provide a degree of fault tolerance for the initial operation. The residue codes were chosen because they are separable, that is the information word and checkbits exist as two separate entities, and because they offer the following advantages when used in a VLSI environment.

1. The checking hardware can be implemented almost independently of any logic which will manipulate the information bits, simplifying the incorporation of this technique into existing hardward libraries.
2. Unlike the non-separable codes (e.g. M-out of-N codes) (Anderson and Metze, 1973) they do not require decoders. The decoders can introduce extra delays and increase the number of different modules required.

The suitability of the residue codes to VLSI designs will be demonstrated by applying the technique to the following functions. In general most of the operations that can be performed at the low level, as opposed to the global view, can be broken down into two groups:

1. Arithemetic operations – addition, subtraction, multiplication and shift operations.
2. Logical operations – AND, OR and Exclusive-OR.

It is well documented (Peterson and Rabin, 1959) that the only code that is capable of detecting faults in logical operations, implemented using random

The design is assumed to be produced in 2 μm CMOS

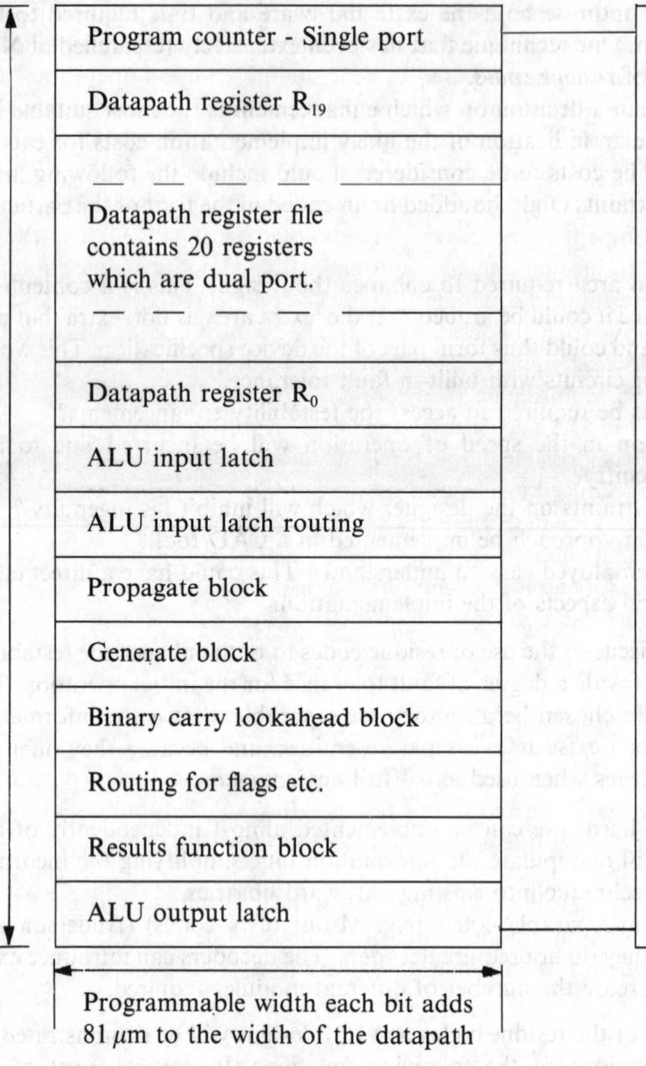

Program counter - Single port	Control Logic Block
Datapath register R_{19}	
Datapath register file contains 20 registers which are dual port.	
Datapath register R_0	
ALU input latch	
ALU input latch routing	
Propagate block	
Generate block	
Binary carry lookahead block	
Routing for flags etc.	
Results function block	
ALU output latch	

Programmable width each bit adds
81 μm to the width of the datapath

Fig. 15.1 *Block diagram of the FRISP datapath*

logic, is straight duplication (see Appendix A for a short-form proof). However, all is not lost, since in certain codes simply by duplicating one logical operation, usually AND, it is possible to generate the checkbits associated with other logical operations. Another consideration that can be uniquely applied to integrated circuits is the method used to generate most logical operations. In a VLSI circuit the use of random logic is to be discouraged for several reasons – mainly the difficulty of designing random logic reliably and quickly, also the possible non-regularity of the structure and the difficulty of automating this form of logic synthesis. One form of logic structure that is recommended for use in a VLSI environment is the programmable logic array (PLA). The PLA is regular, easy to produce reliably and automatically from a few logic equations and can be made testable in a manner that does not depend on the function. The PLA structure allows logical operations to be checked using code bits as will be shown in the following sections. Therefore by constraining the designer to use PLAs it is possible to automate the production of the check PLA.

Each individual logical and arithmetic operation will be discussed. However, to illustrate all of the possible design methods in one coherent theme the design of a fault-tolerant reduced instruction set processor (FRISP) will be presented. The overall layout of the design is shown in Figure 15.1 along with the relevant dimensions. The chip itself is designed to be fabricated in $2\,\mu m$ double layer metal CMOS and will employ certain physical as well as geometric design rules in order to optimise the performance of the chosen coding scheme.

15.2 Residue codes

The residue number system and its attributes are well documented (Garner, 1959) and will not be discussed in this report. However, before presenting two methods of using the residue code, a few definitions will be given:

Definition 1. N is a separable code if and only if the information and check digits can be obtained without any decoding of the codeword, e.g. N is of the form $N = \{X : IC\}$, where I = information digits and C = check digits.

Definition 2. N is called a residue code with check base b if and only if:

1. N is a separable code
2. $N = \{X : X = IC$, given that C is the binary representation of the residue modulo b of I considered as an integer number$\}$.

Definition 3. N is called a *low-cost residue code* if and only if N is a residue code with check base $b = 2^p - 1$, $p \geqslant 2$, where p is the number of checkbits.

In the separate residue codes the number N is coded as a pair $(N, |N|_b)$, where $|N|_b$ is the check symbol of the number N, $|N|_b$ being defined as the least non-negative integer congruent to N modulo b with b the base of the residue

Table 15.1: Common Arithmetic Operations

Operation	Output	Output residue
ADD	$N_1 + N_2$	$\||N_1|_b + |N_2|_b\|_b$
SUBTRACT	$N_1 - N_2$	$\||N_1|_b - |N_2|_b\|_b$
COMPLEMENT	$- N_1$	$- \||N_1|_b\|_b$
MULTIPLY	$N_1 * N_2$	$\||N_1|_b * |N_2|_b\|_b$

code. For example if $N = 31$ and $b = 15$ (a LCR with $p = 4$) then $|N|_b = |31|_{15} = 1$. The ability of the LCR codes to check arithmetic operations has been known for some time and proofs for addition and multiplication can be found in Breuer and Friedman (1977). The sum of the residues modulo b is equal to the modulus of the sum of the information numbers. Similarly the product of the checkbits modulo b is equal to the modulus of the product of the information numbers. Although addition and multiplication are simple arithmetic operations it is possible to extend the technique to subtraction and complement operation (Rao, 1974). Table 15.1 shows the more commonly used arithmetic operations with their corresponding residue outputs. As residue codes are closed under arithmetic operations, they are very simple to use for the functions shown in the table. However, the problem arises when using logical operations, since the residue codes are not closed under bitwise operations such as AND, OR and Exclusive-OR. This situation can be resolved by using the simple theorem which relates arithmetic and logical operations stated below (Monteiro and Rao, 1972).

Theorem 1. If N_1 and N_2 are two n-bit binary vectors, then:

$$N_1 + N_2 = (N_1.N_2) + (N_1 \cup N_2)$$

and

$$N_1 \oplus N_2 = (N_1 + N_2) - 2(N_1.N_2)$$

where the operations '.', '$\cup$' and '$\oplus$' are used to denote the bit by bit logical AND, OR and Exclusive-OR respectively. This theorem can be proved by showing that the relationship holds for 1 bit and then extending it to all n-bits.

From these relationships expressions can be derived for the output residue of the $\cup$ and $\oplus$ operations:

$$|N_1 \cup N_2|_b = \||N_1|_b + |N_2|_b - |N_1.N_2|_b\|_b$$

$$|N_1 \oplus N_2|_b = \||N_1|_b + |N_2|_b - 2|N_1.N_2|_b\|_b$$

These operations are shown in Table 15.2.

So far no indication has been given as to the value of b, although it has been indicated that a low cost residue code must be chosen for optimum performance.

Table 15.2: Common Logical Operations

Operation	Output	Output residue
AND	$N_1 \cdot N_2$	$\lvert N_1 \cdot N_2 \rvert_b$
OR	$N_1 \cup N_2$	$\lvert \lvert N_1 \rvert_b + \lvert N_2 \rvert_b - \lvert N_1 \cdot N_2 \rvert_b \rvert_b$
EXCLUSIVE-OR	$N_1 \oplus N_2$	$\lvert \lvert N_1 \rvert_b + \lvert N_2 \rvert_b - 2 \lvert N_1 \cdot N_2 \rvert_b \rvert_b$

Therefore the following criteria can be used as a set of guidelines to make an informed decision on the choice of residue base.

1. In general the size and complexity of the checking hardware increases sharply as the base increases.
2. As the base increases more bits are required to represent the residue. Although this does mean that fewer, but larger, residue generation blocks are required. It does also mean that the number of pins required on the chip will increase in order to allow the residue bits to leave the chip.
3. The ability of the code to detect errors does not improve significantly as the base increases.

Therefore in order to keep the extra hardware required to a minimum and to keep a reasonable error detection ability a base 3 LCR code ($b = 3, p = 2$) is used. This also has the advantage that a simplication can be made to the residue equations for checking logical operations as shown below.
Let

$$N_3' = \lvert N_1 . N_2 \rvert_3$$

then

$$\lvert N_1 \cup N_2 \rvert_3 = \lvert N_1 + N_2 + \bar{N}_3' \rvert_3$$

and

$$\lvert N_1 \oplus N_2 \rvert_3 = \lvert N_1 + N_2 + N_3' \rvert_3$$

The hardware required to implement all of these operations will be discussed in the following sections.

15.3 Hardware required to implement residue coded designs

To allow a designer to produce circuits based on the residue code scheme it is necessary to produce certain fundamental cells for incorporation into the library. The cells required are as follows:

1. Comparator – this block is required to compare the generated and the calculated residues. The output is in the form of a 1-out of-2 code.
2. Two-rail encoder – this block is required if more than a single 1-out of-2 code

line needs to be taken off chip. This allows only two output pins to be allocated to the error indication signal, if desired. This block can be used to limit the number of pins required on the chip, however if pin limitations are not a problem then the fault resolution can be increased by taking all the fault signals from the chip via individual pins.

3. Residue adder – this block is required for the addition of residues and in the generation of the residue from the information bits. The code chosen will require a mod 3 adder.

4. Residue multiplier – this block is required to produce the product of two residues. This will require a mod 3 multiplier.

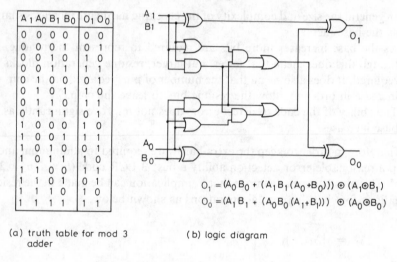

A_1	A_0	B_1	B_0	O_1	O_0
0	0	0	0	0	0
0	0	0	1	0	1
0	0	1	0	1	0
0	0	1	1	1	1
0	1	0	0	0	1
0	1	0	1	1	0
0	1	1	0	1	1
0	1	1	1	0	1
1	0	0	0	1	0
1	0	0	1	1	1
1	0	1	0	0	1
1	0	1	1	1	0
1	1	0	0	1	1
1	1	0	1	0	1
1	1	1	0	1	0
1	1	1	1	1	1

$$O_1 = (A_0 B_0 + (A_1 B_1 (A_0 + B_0))) \oplus (A_1 \oplus B_1)$$
$$O_0 = (A_1 B_1 + (A_0 B_0 (A_1 + B_1))) \oplus (A_0 \oplus B_0)$$

(a) truth table for mod 3 adder   (b) logic diagram

Fig. 15.2a *Mod 3 adder block*

The first two blocks can be used in any design requiring the comparison of data and the compression of two-rail code error information data before being taken off the chip. The final blocks are specific to the residue checking scheme, the form of these blocks will ultimately depend on the chosen value of *b*. The mod 3 adders to be used in this discussion can be produced from a normal 2-bit adder with end around carry (Ashjaee and Reddy, 1976). As the circuit will be used in a multilevel tree structure, a ripple carry design is not suitable as it will slow down the complete system. To overcome this problem and speed up the operation a carry look-ahead arrangement has been chosen. This circuit is larger than the ripple carry scheme, but this is felt to be a reasonable trade off in order to speed up the overall checking operation. The adder produced by this route will produce a '11' output residue in some cases when a '00' residue output would normally be expected. For example, if the value input to the adder were $3 + 3 = 6 \rightarrow |6|_3 = 0$ then this adder would produce 3. Although this anomaly may appear to be a problem, in fact it is a bonus, as it allows the full range of residue bits (00, 01, 10 and 11) to be applied to the comparator. This increases the probability of detecting single faults in the comparator earlier, thus avoiding the problem of multiple errors affecting the results. The circuit diagrams for all these blocks are shown in Figure 15.2. The residue bits can be easily generated from the information bits by using a tree of mod b adders.

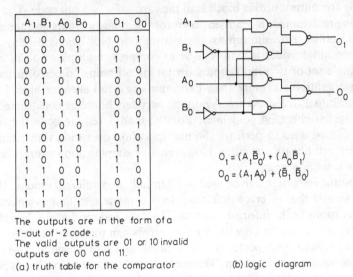

A_1	B_1	A_0	B_0	O_1	O_0
0	0	0	0	0	1
0	0	0	1	0	0
0	0	1	0	1	1
0	0	1	1	1	0
0	1	0	0	0	0
0	1	0	1	0	0
0	1	1	0	0	0
0	1	1	1	0	0
1	0	0	0	1	1
1	0	0	1	0	0
1	0	1	0	1	1
1	0	1	1	1	1
1	1	0	0	1	0
1	1	0	1	0	0
1	1	1	0	1	1
1	1	1	1	0	1

$$O_1 = (A_1 \bar{B}_0) + (A_0 \bar{B}_0)$$
$$O_0 = (A_1 A_0) + (\bar{B}_1 \bar{B}_0)$$

The outputs are in the form of a 1-out of-2 code.
The valid outputs are 01 or 10 invalid outputs are 00 and 11.

(a) truth table for the comparator   (b) logic diagram

Fig. 15.2b *Mod comparator*

A_1	A_0	B_1	B_0	O_1	O_0
0	0	0	0	0	0
0	0	0	1	0	0
0	0	1	0	0	0
0	0	1	1	0	0
0	1	0	0	0	0
0	1	0	1	0	1
0	1	1	0	1	0
0	1	1	1	1	1
1	0	0	0	0	0
1	0	0	1	1	0
1	0	1	0	0	1
1	0	1	1	1	1
1	1	0	0	0	0
1	1	0	1	1	1
1	1	1	0	1	1
1	1	1	1	1	1

$$O_0 = A_0 B_0 + A_1 B_1$$
$$O_1 = A_1 B_0 + A_0 B_1$$

(a) truth table for the two rail encoder   (b) logic diagram

Fig. 15.2c *Two rail encoder.*

The comparator is the most important part of the circuit design as it is used to compare the expected and generated residues for equality. Consequently any faults occurring in this block must be detected immediately or not produce invalid results – that is, the comparator must be designed to be totally self-checking. The design can be simplified if the set of input patterns that are received during normal operation is as large as possible. This is simple to achieve, since, as mentioned above, the residues produced will cover all possible patterns. The output of this block is in the form of a two-rail code. The 01 and 10 values are assumed to represent correct – i.e., valid - inputs and the outputs 00 and 11 indicate invalid inputs.

The two-rail encoder is simply used to compress multiple two-rail outputs into a single set of two-rail outputs for taking off chip. To perform the compression operation for more than two inputs two-rail encoder blocks can be assembled into a tree structure. Again the encoder should be constructed along the totally self checking guidelines. For the mod 3 residue code the two-rail encoder can be used to perform the multiplication operation thus simplifying the overall cell library structure. However, this simplification cannot be made for other residue bases.

The residue code itself can be used in essentially two different modes. The first mode is simply that of error detection. In this mode residue bits undergo the same operations as the information bits. The data and residue are then checked after every operation for consistency, any errors can therefore be detected by a mismatch between the expected (i.e., generated) and actual residue. This is fairly straightforward error detection. The second mode of operation involves using two different low-cost residue codes. Both residue codes are now used to check the operations, but in this case if a single fault occurs, it is possible to correct

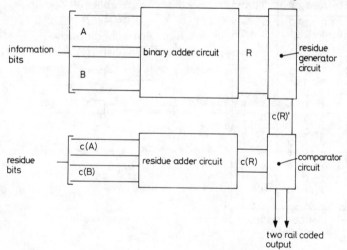

Fig. 15.3 *Generalised residue checking scheme for an adder circuit.*
c(A), c(B) and c(R) are the residue bits associated with the information bits A, B and R respectively.

the error. This scheme can use the same blocks as the first error detection scheme, however it requires a range of mod *b* adder cells as opposed to the single type of adder cells in the first scheme. Therefore by using residue codes the possibility of performing error detection and/or correction with the same hardware is a reality. The next section will discuss these two schemes in more detail and present a few examples.

15.4 The single-residue and biresidue error detection/correction scheme

The single residue used for checking hardware is perhaps the simplest form of error detection. The organisation of the circuits used to produce a design that is checked using a residue code is shown in Figure 15.3. From this diagram it is quite easy to see the relationships between each of the blocks discussed in the previous section. In the example a simple adder is considered to be the checked hardware. However, any blocks can replace the adder and the mod b adder. The only condition that needs to be satisfied is that the 'residue block' performs the same operation on the residue as the 'function block' performs on the data, thus producing a valid residue. By using this scheme it is only possible to detect errors and not perform any correction operations.

In order to correct data using the residue technique it is necessary to use two residues, the codeword being encoded as the triple $(N, |N|_B, |N|_C)$, where B and C are the bases of the two code bits of the information N. Again the low-cost residue code can be used as the check bases for the check bits. In this scheme both the bases $B = (2^b - 1)$ and $C = (2^c - 1)$ must be relatively prime in order to optimise the code, also the number of information bits $k = b.c$, taking these conditions into account ensures that closure under all operations is maintained; proofs of these lemmas can be found in Rao (1974). Since we have stipulated that $b.c = k$, this allows distinct error conditions to be detected and corrected. If a syndrome corresponding to the codeword N is defined as the pair $\langle s_1, s_2 \rangle$, where $s_1 = (|N - |N|_B|_B)$ and $s_2 = (|N - |N|_C|_C)$. From the definitions

Table 15.3 *Biresidue syndromes to correct a 12-bit number*

j	syndrome of $+2^j$ $\langle s_1, s_2 \rangle$	syndrome of -2^j $\langle s_1, s_2 \rangle$
0	(1,1)	(6,14)
1	(2,2)	(5,13)
2	(4,4)	(3,11)
3	(1,8)	(6,7)
4	(2,1)	(5,14)
5	(4,2)	(3,13)
6	(1,4)	(6,11)
7	(2,8)	(5,7)
8	(4,1)	(3,14)
9	(1,2)	(6,13)
10	(2,4)	(5,11)
11	(4,8)	(3,7)

of s_1 and s_2 if there is no single error in the codeword, then $s_1 = 0$ and $s_2 = 0$. Since we are assuming that only single-fault conditions will occur, then only errors of the form $\pm 2^j$ ($j = 0, 1, \ldots, k - 1$) will be produced. Therefore for these error conditions $s_1 = \pm 2^j \bmod (2^b - 1)$ and $s_2 = \pm 2^j \bmod (2^c - 1)$. It can be shown that the syndromes for all these errors are distinct. Therefore when $s_1 \neq 0$ and $s_2 \neq 0$, there is an error in the processing element. If $s_1 \neq 0$ and $s_2 = 0$, then there must be an error in the checker producing the syndrome s_1, the converse is also true for s_2. If one of the residues is incorrect, then the

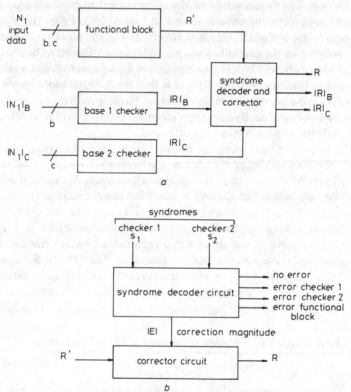

Fig. 15.4 *Example of the Biresidue correction scheme: (a) Checker and functional block architecture, (b) syndrome decoder and corrector circuits.*

value can be recomputed from the information bits. Since the syndromes produced by the error checkers are distinct, then it is possible to correct all single errors. This can be illustrated by the following example.

Example 1. If $B = 7 = 2^3 - 1$ and $C = 15 = 2^4 - 1$, this will give $k = 3.4 = 12$ allowing the values $m = 2^{12} - 1$. Table 15.3 below gives the error syndromes for all values of $j = 0, 1, \ldots, 11$.

The detection capabilities of the biresidue code are identical to those of the single residue scheme. Figure 15.4 shows the details of a processor that might be used with the biresidue code. Basically the hardware required is similar to that required for the single residue scheme, except of course that two different residue generation circuits are required. A syndrome decoder is also required to produce the error magnitude and indicate the type of error that has occurred. This will obviously increase the complexity of the hardware quite considerably. It also means that the number of pins on the chip will increase. One further problem introduced by the biresidue scheme is that the number of bits in the information word must be equal to the product of the number of bits in the two residue bases for optimum performance. This means that it is not possible to have information bits which form powers of two. For example, the closest pair of low-cost residue codes to form a 32-bit number are 5 and 6 (i.e., bases 31 and 63), which will give the capability to check information with 30 bits. therefore the usefulness of this code might be in checking schemes with processors of an odd number of information bits. This also applies to whether or not the residue bits are corrected when they are determined to be in error. The following section examines the costs of implementing these two schemes in typical VLSI structures.

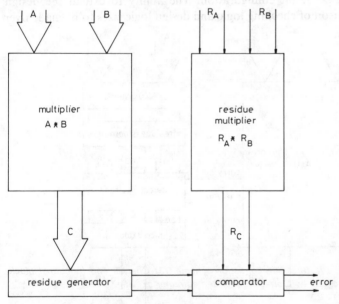

Fig. 15.5 *Layout of a self-checking multiplier.*

15.5 Implementation costs for the residue codes

15.5.1 Applications of residue coding
To demonstrate the applicability of the residue coding technique to a wide range of functions the following circuits are chosen as examples:

- $N \times N$-bit multiplier (Yung and Allen, 1984)
- N-bit ALU

- N-bit register (memory elements)
- PLA structures (Mead and Conway, 1980)

15.5.2 Multiplier

As multiplication is an arithmetic operation it is possible to derive a relationship between the input and output of the device, as demonstrated in the previous section. Figure 15.5 shows an example layout which can be used to check the multiplication operation. In the diagram A and B are the inputs and C is the product $A \times B$. The residues of A, B and C are given by R_a, R_b and R_c respectively; $R_c = |R_a \times R_b|_3$. Once the output residue has been generated, it is compared with the residue generated from the actual multiplication operation itself.

The checking hardware was applied to an $N \times N$-bit multiplier (Yung and Allen, 1984), which is constructed from two basic building blocks, a 2×2-bit multiplier, and a full adder. The use of such a simple arrangement allows the extension of the multiplier from its minimum 4×4-bit multiplication configuration to a 32×32-bit configuration. The ability to extend the design allows the comparison of checking logic and design logic sizes to be made over

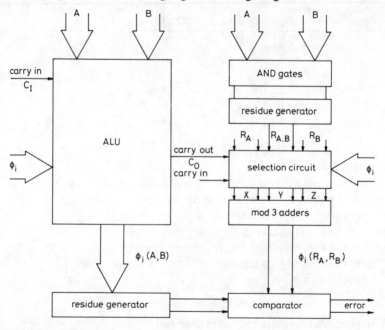

Fig. 15.6 *Organisation of a self-checking ALU.*

a wide range of input bit lengths. As the checking logic is independent of the implementation of the function, it can be incorporated into any device which performs a multiplication operation. A biresidue check on the multiplier design would simply require the addition of extra hardware to cover two residues as opposed to the single residue. However, difficulties arise when attempting to

Table 15.4: Residues required to predict the output from an ALU

Mnemonic	Bit number		
	1	2	3
Addition $A + B$	R_A	R_B	$C_o C_I$
Subtract $A - B$	R_A	$\overline{R_B}$	$C_o C_I$
Increment $A\ A + 1$	R_A	00	$C_o C_I$
Increment $B\ B + 1$	00	R_B	$C_o C_I$
Complement $A\ \bar{A}$	$\overline{R_A}$	00	$C_o C_I$
Complement $B\ \bar{B}$	00	$\overline{R_B}$	$C_o C_I$
Decrement $A\ A - 1$	R_A	00	$C_o C_I$
Decrement $B\ B - 1$	00	R_B	$C_o C_I$
Clear 0	00	00	$C_o C_I$
Subtract $B - A$	$\overline{R_A}$	R_B	$C_o C_I$
AND $A.B$	00	00	$R_{A.B}$
OR $A \cup B$	R_A	R_B	$R_{A.B}$
Exclusive- OR $A \oplus B$	R_A	R_B	$R_{A.B}$

R_A = Residue of the A input
R_B = Residue of the B input
$R_{A.B}$ = Residue generated from the duplicate AND operation
$C_o C_I$ = Carry out C_o and Carry in C_I
The output residue is formed by adding the three residues together in a 'mod b' adder. For some
e.g. decrement and increment operations the carry in is usually forced to a 1

correct the output, since the two inputs would require larger code bits than necessary to cope with the expansion of the bits to the new range at the output. Consequently a multiplier would present severe difficulties if included in any biresidue checking scheme.

15.5.3 ALU
The main problem with applying residue checking to ALU designs is that they can perform both logical and arithmetic operations. Consequently, extra provision has to be made to check the logical operations. Figure 15.6 shows an example of how the checking logic could be applied to an ALU. Table 15.4 shows the required residues for the operation of a typical ALU. The main differences between incorporating residue checking into an ALU and a multiplier, which can only perform an arithmetic operation, are:

1. The inclusion of logic to produce the mod 3 residue of the 'AND' operation ($|N_1 . N_2|_3$). This is used in the generation of residues for other logical operations such as 'OR' and 'Exclusive-OR'.
2. The selection circuit to choose the residues to be used in the final generated residue. This is necessary because the ALU performs many different operations. Table 15.4 shows the residues required to predict the output residues for various ALU operations. The need for the carry-in and carry-out in the selector circuit will be discussed in 15.8.2. The output of the ALU can be checked in a similar way to the multiplier.

The ALU to which this technique was applied, was expandable for 4 bits to 32 bits per input, allowing the checking logic to be applied to a wide range of input bit lengths. With slight modifications to the selector circuit (to allow for different functions) the design shown can be applied to any ALU. If biresidue codes were to be used, then two residue generating ALUs would be required both of different bases, along with independent residue generation blocks.

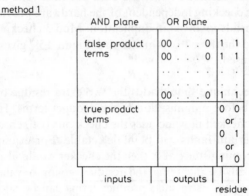

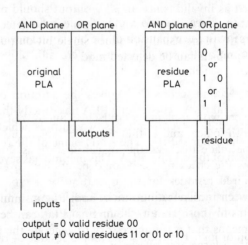

Fig. 15.7 *Examples of PLA layouts.*

15.5.4 PLA

The residue checking technique was applied to a range of PLA structures of various complexity for example, the traffic-light controller described in Mead and Conway (1980), a DMA controller and an electronic combination-lock controller. The problem with trying to produce a standard technique to check

PLAs is that PLAs do not perform standard functions, whereas ALUs and multipliers always perform basically the same operations, the internal architecture being irrelevant to the checking problem. Consequently in arithmetic operations the checker is essentially independent of the hardware to be checked, and therefore since the two systems (checker and arithmetic hardware) are separate, unchecked common hardware does not exist. The aim in checking a PLA is to provide checking independent of the hardware providing the outputs, without the overheads involved in duplication. Mod 3 checking was applied to by two methods to the three PLAs mentioned, Figure 15.7 gives a block diagram for each method used.

Method 1. This technique simply adds the two extra residue bits on to the side of the OR plane before minimisation of the product terms. However, because the system being checked now includes the check bits (i.e., checker hardware) it is possible that certain faults will be undetectable. For example, if a product term failed by becoming stuck-at 0, then the checker would also fail because the output would be all 0s including the residue, giving a valid codeword. To overcome this problem extra false product terms can be added to the AND plane. The false product terms are those terms which are associated with an all-0 output and are usually ignored in PLA designs. The false product terms are associated with a residue of 11. Thus if a true product term (i.e, a term which produces at least one true output) were to fail as previously described, its output could be treated as invalid, since an all-0 output should produce a 11 residue instead of 00. Most errors in the AND plane can be detected by this method. Errors in the 'OR' plane usually produce single bit output errors (Wang and Avizienis, 1979) which can be detected mod 3.

Method 2. The second technique involves the addition of a second, residue PLA, to the original PLA. The residue PLA has exactly the same inputs as the PLA being checked; however it only has two outputs, which are the residues of the outputs of the original PLA. The AND plane of the residue PLA can be a minimised version of the original PLA. This method has several advantages over the first method:

1. Since an error is unlikely to occur in both PLAs simultaneously, identical product terms to those used in the original PLA can be used in the residue PLA. This means that it is unnecessary to generate the false product terms, which require a large area overhead.
2. As the residue PLA is independent of the PLA being checked, addition of the checker hardware to the system does not affect the checked PLA, consequently the intended operating characteristics will not be affected.

This checking method has similar error-detection characteristics to Method 1, as errors are highly unlikely to occur simultaneously in both PLAs, and the detection characteristics are mainly due to the properties of the code.

In both the cases presented the residue generation circuit must be capable of checking the all-0 output. The two methods above present slightly different problems. The second method allows an all-0 output with a corresponding 00 residue, and here, if a 00 residue is normally produce by a valid non-zero output it would not be possible to detect a faulty product term, for this reason a 00 residue should be avoided, if possible, for non-zero outputs. There are two possible ways of achieving this:

1. Extend the 'OR' plane by one bit, to allow only 01 or 10 residues, and increase the checker size.
2. Use a 11 output for these terms instead.

Method 2 is preferred for two reasons: it does not affect the original PLA and it fits in well with the residue generation and checker hardware already proposed. The second method of implementing the 00 residue in the PLAs is also to be prefered since it will easily fit in with the scheme proposed. However, in the next section methods 2 and 2 will be used to show how the size of the PLA can be expected to increase with their use. Again the biresidue scheme can be used with this technique, however optimum performance will not be easy to achieve since the number of bits in a PLA can vary.

15.5.5 Memory elements
As a memory element is only a storage device (i.e. it performs no function between the input and output), it is only necessary to store the residue or residues associated with a particular bit sequence. This requires the addition of two extra blocks of storage (for mod 3 checking) to the word storage. The biresidue code will require enough extra storage to maintain the residue throughout the system.

15.6 Properties of the residue codes

As already discussed, two of the important factors governing the choice of test method in a particular application are these:

1. The cost in terms of silicon area required to implement the chosen test method when compared to the design without the test structures.
2. The time penalties imposed on the design by any parts of the test hardware.

The extra area required by the residue checking technique was therefore calculated in order to compare it with other common test methods. The results will be discussed in this section. The increase in area, resulting from the additonal mod 3 residue checking hardware in the multiplier and ALU structures is shown graphically in Figure 15.8, where only extra area taken by the test hardware, and not the interconnection of the hardware, has been considered.

The multiplier shows the best performance as the number of bits in the design increases. This is because the residue prediction hardware size remains almost constant, although the checker circuits do increase in size. The ALU does not show such a good performance, partly because of the increase in the AND residue prediction hardware as the number of bits grows. No calculations were performed for the multiplier circuit with the biresidue code since it was felt to be an impractical proposition. The calculations for the biresidue scheme applied to the ALU are shown in Table 15.5.

Table 15.6 shows the results obtained from the three types of PLA to which mod 3 residue checking was added. Generally, it was found that the more

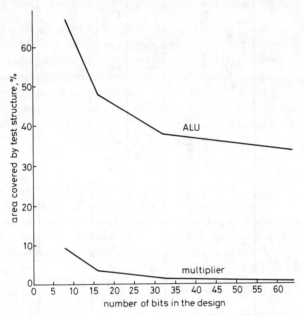

Fig. 15.8 *Mod 3 residue check.*

overlap between product terms (i.e., the fewer inputs to the product terms) the better the minimisation of the false product terms and the residue PLA AND plane. The residue PLA method, although requiring more silicon area, was generally easier to generate by software means and produced a more constant

Table 15.5: Increase in area when adding the Biresidue code to an ALU

Biresidue error detection on the ALU			
No. Bits	Size of ALU $(\mu m)^2$	Size of check ALU $(\mu m)^2$	Size of checkers $(\mu m)^2$	Percent increase %
12	1.49×10^6	8.0×10^5	1.0×10^6	123
20	2.48×10^6	1.0×10^6	1.7×10^6	111
30	3.72×10^6	1.3×10^6	2.9×10^6	104
56	6.9×10^6	1.7×10^6	4.8×10^6	96

Table 15.6: Size increases in PLA structures

PLA name	No. inputs	No. outputs	No. product terms	Method 1 Extra			Method 2 New PLA		
				Size increase %	Outputs	Product terms	Extra area required %	Inputs	Outputs	Product terms
Traffic light Controller	5	8	10	8	2	0	73	5	2	9
DMA Controller	12	12	26	27	2	8	67	12	2	22
Electronic combinational Lock control	15	6	18	213	2	50	93	15	2	20

size increase, allowing size increases for other residue PLAs to be predicted. The checking hardware associated with the PLAs was not included in any of the estimates, but could be expected to increase the size by an extra 10–20%. No biresidue structures were produced since the practicalities of applying this technique to PLAs with varying numbers of outputs were considered to be insurmountable.

It is difficult to produce estimates for the size of the checking logic required by memory elements, since they could be used in conjunction with other devices (e.g., ALUs, which are capable of performing the check operation themselves); or alternatively, the memory elements could be used in a larger block, such as a RAM, in which case checking could be applied to the block as a whole, allowing a complete RAM chip to be checked by one piece of checking hardware. For a 16-bit register, an increase of about 14% can be expected in the register size to store the 2 bits required by the mod 3 residue, ignoring the overheads of the checking hardware – i.e., residue generators etc.

As a check on the abilities of the mod 3 code to detect faults, a small-scale simulation was made on the 4×4-bit multiplier, discussed earlier. In the simulation the most probable single bit errors and a range of multiple bit errors were introduced. The overall probability that an error would be detected was found to be 99.8% for the mod 3 residue code and only 68% for the single bit parity code. Most of the faults were introduced into the adder part of the circuit, since this is the point at which faults that occur may cause burst errors, especially if the carry is affected; several output bits could then be altered at the same time.

15.7 Miscellaneous operations

This section will discuss a few problems that arise in the application of this mod 3 residue technique to certain arithmetic operations and their possible solutions. The main problems arise because of the use of 2-bit adders with end around carry to generate the residue. This produces a 11 residue for input data divisable by 3, except for the zero datum. Consequently it is, on occasions, necessary to take corrective action to restore the residues to consistent values. The conditions that necessitate the corrective action are the cases when the residue produced is 00 and the actual information bits are not zero and when the residue is 11 and the information bits are zero.

Another problem arises because of the residue codes themselves. When performing modulo 2^n arithmetic, as in a normal binary adder, it is possible to produce a carry-out in the $(n+1)$th bit position. The residue generated however will in fact cover all the $(n+1)$ bits. Therefore the residue has to be corrected so that it only applies to the n bits of the normal word.

Possible solutions to these problems will be discussed in connection with the one's complement and two's complement number system. The rotate and logical shift operations are also discussed.

15.7.1 One's complement number system

Mod 3 checking can be applied easily to this number system since the carry-out bit is usually added back into the number using an 'end around' carry.

One's complement negation simply involves performing the NOT operation on every bit in the number. This operation can also be performed on the residue to produce the residue of the negated number. However, becaue of the form of the residue encoding implemented, corrective action must be taken whenever a number with residue 11 is negated, if the negation operation does not produce the all zero's output. Alternatively since the one's complement system has two representations of 0 only the all 1's case could be used. This would mean a residue of 11 would never have to be inverted. A few examples of this operation are presented in Figure 15.9, a circuit which can be used to correct the residue is also shown.

		Residue
+ 7	0 1 1 1	0 1
− 7	1 0 0 0	1 0

However,   Use Z and C bits
| − 6 | 1 0 0 1 | 1 1 | from ALU as shown below
| + 6 | 0 1 1 0 | 0 0 - Incorrect | to produce correct
| | | 1 1 - Correct | Residue
| | | 0_1 0_0 |

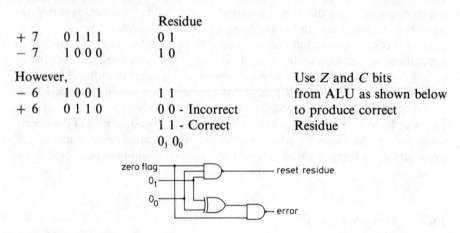

Fig. 15.9 *One's complement negation.*

+ 6	0 1 1 0	1 1		+ 6	0 1 1 0	1 1
-----	---------	-----		-----------	---------	-----
+ 1	0 0 0 1	0 1		+ 4	0 1 0 0	0 1
+ 7	0 1 1 1	0 1		(+ 10) − 5	1 0 1 0	0 1

Overflowed. Residue still valid

− 3	1 1 0 0	1 1
− 4	1 0 1 1	1 0
	0 1 1 1	
1 ⟶ 1 ← End around carry
− 7	1 0 0 0	1 0

Fig. 15.10 *One's complement addition.*

The addition of one's complement numbers is accomplished in the normal way. The residues are simply added together using the mod 3 adder discussed earlier. The final residue of the sum produced by the addition operation. No further corrective action is required. Figure 15.10 shows a few examples of this operation.

Subtraction can be performed as normal. The normal sequence of events is to negate the subtrahend and then to add this number to the minuend. The residues can therefore undergo the same sort of operation, since they will be negated and then added. The additions can be performed by a mod 3 adder. Figure 15.11 shows a few examples of this operation.

15.7.2 Two's complement number system

The application of mod 3 checking to the two's complement number system is not as simple as the description for the one's complement number system. The main problem is the generation of a carry out from the most significant bit, unlike the one's complement system the carry is not incorporated back into the number. Therefore in a real system information will be lost, namely the carry-bit, and the residue will no longer be correct, as it applies to the complete word including the carry bit. There are a number of possible solutions to this problem.

```
  + 7      0 1 1 1    0 1                − 4      1 0 1 1    1 0
+ (− 4)    1 0 1 1    1 0              + (− 1)    1 1 1 0    1 0
           0 0 1 0                                1 0 0 1

            1 ──→ 1                               1 ──→ 1
  + 3      0 0 1 1    1 1                − 5      1 0 1 0    0 1
```

Fig. 15.11 *One's complement subtraction.*

The simplest solution is to regenerate the residue from the data whenever a carry occurs. However this will break down the independence of the two operations and may make the system unreliable.

The technique chosen to solve this problem relies on information that is usually available in the adder – that is, the carry bit itself and the zero flag. The corrected residue can be obtained by subtracting the carry bit from the predicted residue. The result produced will be correct providing the answer is not 0. When the result is 0 the status of the zero bit must be examined. If the zero bit is true then the residue may or may not be correct because of the dual representation used for the zero residue. If a zero residue should have resulted then the final residue could be either 00 or 11, but it will not be 01 or 10. Therefore if the zero flag is true and the resulting residue is 01 or 10, an error has occurred and can be flagged. Otherwise the residue can be cleared to zero for the zero datum, providing the necessary corrective action. Of course any corrective action required for the carry-bit should be performed first. Therefore the zero flag bit is also checked. The carry-bit is also checked, since if it is faulty it would cause an error of $\pm 2^n$ in the residue (Avizienis, 1971), which is easily detected by the mod 3 residue code. The error would be detected when the comparison is made between the residue and the residue generated from the information bits. Figure 15.12 illustrates the technique.

The method used to negate two's complement numbers is to invert all the bits

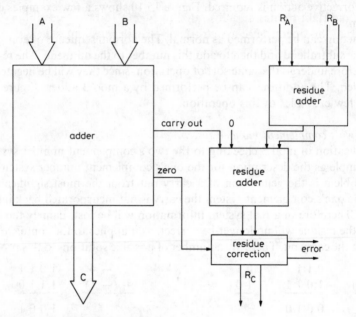

Fig. 15.12 *Correcting the residue using the carry bit.*

		Residue			Residue
+ 1	1 1 1 0	0 1	0	0 0 0 0	0 0
NOT	0 0 0 1	1 0		1 1 1 1	1 1
Add 1	1	0 1		1	0 1
− 1	1 1 1 1	1 1	1̄	0 0 0 0	0 1
				+ 10 →	1 1

C = 1

Z = True ⟶   0 0

Fig. 15.13 *Two's complement negation.*

in the number and then add 1 to the result. This operation can also be performed on the residue. The residue is first complemented and a constant value of 01 is added. Figure 15.13 illustrates this method for the corrective action discussed above.

The addition operation, illustrated in Figure 15.14 additon is performed in the normal way. The two residues of the numbers are added and the corrective action taken. The subtraction operation is illustrated in Figure 15.15.

The operations described for the two number systems are the most basic available. Therefore further operations could be performed by using these basic techniques. Consequently these more advanced operations will also be checked satisfactorily, since they use operations which are themselves checked. Again these techniques can be applied to the biresidue scheme.

15.7.3 Rotate operation

The left and right rotates are illustrated in Figure 15.16. Since all the bits are preserved in this operation, the residue will remain constant and can simply be rotated left or right along with the number. One bit shifts are shown in the example.

```
+ 6        0 1 1 0      1 1      + 6      0 1 1 0      1 1
+ 1        0 0 0 1      0 1      + 4      0 1 0 0      0 1
-----      -------      ---   (10)- 6     1 0 1 0      0 1
+ 7        0 1 1 1      0 1
                                Overflowed. Residue correct


- 3        1 1 0 1      0 1      - 8      1 0 0 0      1 0
- 4        1 1 0 0      1 1      - 8      1 0 0 0      1 0
-----    1 1 0 0 1      0 1            1 0 0 0 0      0 1
Subtract carry ───→     1 0                           1 0
-----      -------      ---   (- 16)  0   0 0 0 0      1 1
- 7        1 0 0 1      1 1
                                       Z = True ───→   00

                                          0 0 0 0      0 0
```

Fig. 15.14 *Two's complement addition.*

```
    + 7     0 1 1 1       0 1         + 7     0 1 1 1      0 1
+               +
    - 4     1 1 0 0       1 1         - 8     1 0 0 0      1 0
          1 0 0 1 1       0 1         - 1     1 1 1 1      1 1
          Subtract carry ───→  1 0
    + 3    ·0 0 1 1       1 1
```

Fig. 15.15 *Two's complement subtraction.*

15.7.4 Logical shifts

Logical shifts are more difficult to check because of the loss of bits as the shift operations take place. Figure 15.17(a) shows the logical shift operation. The circuit in Figure 15.17(b) demonstrates how the loss of these bits can be dealt with, the system uses the same corrective action as the two's complement number system. The left shift operation simply involves rotating the residue left then subtracting the carry-out from this value. The right shift is very similar in that the residue is rotated right and then any carry-out is added to the residue. It is also necessary to clear the residue if the result of a shift produces a zero result, again this can be checked as before.

15.8 Cost of implementing the checking hardware

The area overheads incurred by incorporating both residue checking schemes into the FRISP datapath are shown in Tables 15.7 and 15.8. The circuit is implemented using $2\,\mu$m 5V CMOS technology.

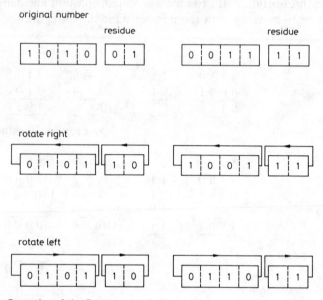

Fig. 15.16 *Examples of the Rotate operation.*
The rotate operations do not need any corrective action on their residues.

The degradation in performance of the system is minimal, the major drawback being the necessity to wait for the carry and zero flags to be generated by the ALU before the residue prediction hardware can complete its operation. This would deteriorate as the number of bits in the design increased, unless steps were taken to speed up the carry operation overall.

It has been reported that the extra hardware required to test a 32-bit CMOS processor chip using scan path is estimated to be between 25% and 40%,

Table 15.7: Increase in area using single residue detection

	Single residue error detection			
No. Bits	Size of datapath $(\mu m)^2$	Size of check ALU $(\mu m)^2$	Size of checkers $(\mu m)^2$	Percent increase %
8	3.16×10^6	1.1×10^6	2.6×10^5	44
16	6.32×10^6	1.1×10^6	6.16×10^5	27
32	1.2×10^7	1.1×10^6	1.32×10^6	20
64	2.52×10^7	1.1×10^6	2.73×10^6	15

Table 15.8: Increase in area using biresidue correction

	Biresidue error correction		
No. Bits	Size of datapath $(\mu m)^2$	Size of check ALU $(\mu m)^2$	Size of checkers $(\mu m)^2$	Percent increase %
12	4.74×10^6	3.6×10^6	1.0×10^6	98
20	7.9×10^6	4.6×10^6	1.7×10^6	81
30	1.2×10^7	5.7×10^6	2.6×10^6	69
56	2.2×10^7	7.7×10^6	4.84×10^6	57

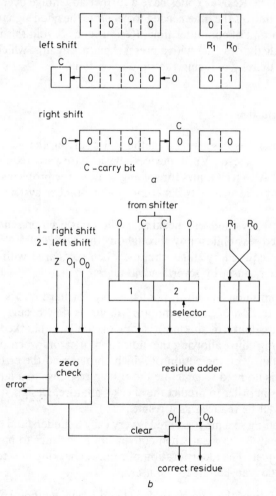

Fig. 15.17 *Logical shift operations.*

increasing to 70% when extra control logic etc. is included. In the case of this datapath because of the large number of registers already present in the design if scan paths or BILBOs were to be used the increase in size might only be of the order of 20%, mainly due to the inclusion of multiplexor blocks in the register for the scan path test method and Exclusive-Or gates with feedback paths if a BILBO method is to be used. However, scan path and BILBO techniques are intended to be employed as a single test of the design either before or after packaging, sometimes they may be used to assist diagnosis of a faulty component. Residue codes are not only intended to provide ease of testing initially but also to continuously monitor the operation of the chip once it is in service. Residue codes have a further advantage over BILBO techniques in that it removes the time needed to generate the good signature which can take a large amount of computer time. Residue codes avoid this in that they essentially provide the designer with a preset signature against which a comparison can be made, thereby reducing the simulation time.

15.9 Conclusions

This chapter has descussed a method of applying the mod 3 low cost residue code to a range of VLSI devices. Basic hardware requirements have been presented. Also techniques for solving some of the problems associated with the one's complement and two's complement number system have been demonstrated.

The mod 3 residue self checking method has been presented allowing the use of a unified test philosophy throughout the design of a VLSI chip. The extra hardware required by this technique when compared with LSSD and BILBO techniques is offset by several advantages:

1. The continuous monitoring of the chip for hard errors once in service.
2. The detection of most temporary faults as they occur.
3. The possibility of linking a series of similarly checked chips to a central 'checking chip', allowing the indication of errors when they occur and more importantly an indication of which chip caused the error.
4. There is no need to generate the system response to a large number of input vectors in order to predict the check signature. Design and testing time costs may well be reduced as a result.
5. The coding scheme presented is very easy to understand and simple to apply. This may be an asset when convincing a designer to incorporate testability techniques. The incorporation of residue checking in a cell based system can be made transparent to the user.

The ability of these codes to detect and report temporary faults as they occur is probably their most significant advantage. As has already been discussed, the number of temporary errors is likely to increase as the scale of integration becomes even greater and component sizes decrease. The importance of continuous checking at little extra hardware cost is therefore a factor which will weigh more heavily in future VLSI designs, and residue codes have been shown to be a cost effective way of providing such checking.

Since extra active area is being used to implement these designs the yield of the device (Sze, 1983) will almost certainly drop. However it may also drop because the scheme might detect faults that were not considered in the original design phase. If an error detecting/correcting scheme is employed, then it would be possible to obtain three classes of chips after inital testing. There would be the chips that are functional and working to specification. Some chips would be non-operational. The third set of chips might function due to the presence of the coding scheme correcting the errors caused by the internal faults, however they would fail if another error occurred. It is therefore possible that the yield of useful devices using this type of coding scheme might be higher than expected. The third class of chips might be used in less demanding applications where their error correcting abilities are not required. A further benefit of employing an information redundant technique hierarchically in a design is that it permits the integrity of the connections between modules in a circuit to be tested, this will be advantageous in the design of WSI systems.

Acknowledgements

We would like to thank Ian Elliot for his work on the FRISP design and Farid Kazi for the mod 3 cell layouts used in the estimation of area overheads.

Appendix A proof of the duplication property for logical operations

For all non-trivial logic operations ($x.y$, $x.\bar{y}$, $\bar{x}.y$, $\bar{x}.\bar{y}$, $x \cup y$, $x \cup \bar{y}$, $\bar{x} \cup y$, $\bar{x} \cup \bar{y}$) except Exclusive-OR and Equivalence then the required error detecting code is no better than duplication.

Consider two n-bit vectors A and B as inputs to a circuit performing a non-trivial logic operation and producing and output vector R. Assume there is also another circuit which performs the same operation on the check bits of A and B, that is, $c(A)$ and $c(B)$ respectively, to produce the check bits of R, namely $c(R)$. Detection of single errors requires that the chosen code is capable of indicating that one of the digits in the output, R, is incorrect. Therefore in this case R and R^1, the faulty output, must differ in one bit position and $c(R) \neq c(R^1)$. If A and A^1 are two n-bit vectors differing in the ith position and E_i is a vector with a 1 in the ith position and 0s in the other positions, then considering the AND operation performed between the two vectors $A.E_i$ and $A^1.E_i$, they will differ in exactly one position. Therefore it would be expected that

$$c(A.E_i) \neq c(A^1.E_i)$$

therefore

$$c(A) \bigstar c(E_i) \neq c(A^1) \bigstar c(E_i)$$

where $\bigstar$ represents the equivalent operation on the check bits. Therefore $c(A) \neq c(A^1)$. Thus for every input vector the check symbol must be distinct. That is the check method can be no better than duplication.

References

ANDERSON, D. A. and METZE, G. (1973): 'Design of totally self-checking check circuits for m-out-of-n codes', *IEEE Trans on computers*, **C-22**, (No.3), pp. 263–8

ASHJAEE, M. J. and REDDY, S. M. (1976): 'On totally self-checking checkers for separable codes', *Proc. 6th Fault Tolerant Computing Symp.* pp. 151–6

AVIZIENIS, A. (1971): 'Arithmetic error codes: Cost and effectiveness studies for applications in digital systems', *IEEE Trans. on Computers*, **C-20**, (No. 11), pp. 1322–31

BREUER, M. A. and FRIEDMAN, A. D. (1977): *'Diagnosis and reliable design of digital systems'* (Pitman, London)

GARNER, H. L. (1959): 'The residue number system', *IRE Trans. on Computers*, **EC-8**, pp. 147–150

FANTINI, F. (1984): 'Reliability problems with VLSI', *Microelectronics and Reliability*, **24**, (No. 2), pp. 275–96

LIU, S. (1984): 'The role of a maintenance processor for a general purpose computer system', *IEEE Trans. on Computers*, **C-33**, (No. 6), pp. 507–17

MANGIR, T. E. and AVIZIENIS, A. (1980): 'Failure modes for VLSI and their effect on chip design', *Proc. 1st IEEE Conference on Circuits and Components*, pp. 685–8

MAY, T. C. (1979): 'Soft errors in VLSI – Present and Future', *Proc. 29th IEEE Conf. on Electronic Components*, pp. 247–56

MEAD, C and CONWAY, L. (1980): *Introduction to VLSI systems*, (Addison-Wesley, Massachusetts)

MONTEIRO, P. and RAO, T. R. N. (1972): 'A residue checker for arithmetic and logical operations', *Proc. 2nd Fault Tolerant Computing Symp.* pp. 8–13

PETERSON, W. W. and RABIN, M. O. (1959): 'On codes for checking logic operations', *IBM J. Res and Dev.*, **3**, pp. 163–8

RAO, T. R. N. (1974): *'Error coding for arithmetic processors'*, (Academic Press, NY)

SAVARIA, Y., RUMIN, N. C., HAYES, J. F. and AGRAWAL, V. K. (1984): 'Characterisation of soft error sources', *Report No. 84–11*, McGill University, VLSI Design Laboratory

SZE, S. M. (1983): *'VLSI Technology'*, (McGraw-Hill, New York)

TASAR, O. and TASAR, V. (1977): 'A study of intermittent faults in digitial computers', *Proc. AFIPS*, **46**, pp. 807–11

TENDOLAKER, N. and SWANN, R. (1982): 'Automated diagnostic methodology for the IBM 3081 processor complex', *IBM J. Res. and Dev.*, **26**, pp. 78–88

WANG, S. L. and AVIZIENIS, A. (1979): 'The design of totally self checking circuits using PLAs', *Proc. 19th Fault Tolerant Computing Sympo.* pp. 173–89

YUNG, H. C. and ALLEN, C. R. (1984): 'Part 1 – VLSI implementation of a hierarchical multiplier', *Proc. IEE*, (Pt. G), **131**, (No. 2), pp. 56–60

PART 3

Wafer-scale integration

An introduction to WSI

Chapter 16

Wafer scale integration

R. A. Evans

16.1 Introduction and motivation for WSI

The integrated circuit, or chip, has been with us for the past 30 years and has been a significant contributor to the economic growth of the industrialised world. In 1959, only a single transistor could be placed on a chip; by 1987, RAM chips could be fabricated containing over 1 million devices. This has been made possible by major improvements in the fields of material growth device processing and lithography. For example, the $25 \, \mu m$ device geometries of the early 1960s have been reduced to less than $1 \, \mu m$ in some current designs.

Integrated circuits begin life as part of a wafer containing perhaps several hundred identical chips. After processing, the chips are tested, and the wafer is then sawn to produce individual silicon chips. The functional chips are then packaged, retested and marketed. Wafer-scale integration (WSI) is the term used to describe the process of using the entire surface of the wafer for implementing a single circuit, rather than sawing the wafer into individual chips.

The earliest known reference to WSI is by Sack *et al.* (1964), who were dealing with 1 inch diameter wafers, although nowadays, wafers of 6 inches in diameter are in common use. Apart from some work in the mid to late 1960s, little significant work was carried out in the field of WSI until the late 1970s, when Trilogy mounted a major development programme. Since 1980 many research groups throughout the world have taken interest in the concept. The recent upsurge of interest is mainly due to the trend towards parallel processing, in which arrays of identical processing elements are used to perform computation tasks very rapidly by partitioning the problem and processing many parts of the problem simultaneously. The use of arrays of identical processing elements brings a major simplification to the task of designing a WSI circuit. Another motivation for WSI is that the rate of growth in chip complexity is not keeping pace with the desire for more and more computing power. This problem is likely to worsen as material limits are approached, and fundamental changes in processing techniques such as lithography and etching are already being required.

In the next section we consider the advantages and disadvantages of WSI, and some of the problems to be addressed before it becomes a standard implementation technique. In 16.4 to 16.6 we classify and review published techniques for configuring arrays of processors which contain faulty elements, and describe the types of switch which are being considered for this task. In 16.7 we briefly describe some of the WSI devices which have been built so far, and in 16.8 we mention two European research programmes on WSI.

16.2 Advantages and disadvantages of WSI

The potential advantages of WSI are similar to those which were achieved by the move from discrete components to integrated circuits. These include the following:

1. Reduced area due to removal of individual packages for chips
2. Reduced weight
3. Reduced power requirements due to removal of most of the interchip drivers
4. Reduced system cost due to removal of individual packaging and assembly costs
5. Increased system reliability due to removal of chip-to-package, package-to-pcb and inter-pcb connectors
6. Possibility of reconfiguring the system in the field after an in-service failure because fault tolerance will be an essential feature of any WSI device

Some of the disadvantages of WSI are as follows:

1. Strategies for tolerating processing defects must be developed and included in the circuit. This increases the area of the device.
2. Development costs of WSI devices are likely to be high.
3. The circuits produced in WSI will be complex and are likely to have a restricted range of applications. This may mean that devices are expensive

From the above list it is clear that considerable risks are involved in WSI. However, many systems could derive significant benefit from WSI and at this stage there is every indication that research will continue.

16.3 Implementation challenges

The main problem to be overcome in the development of any WSI device is the toleration of the processing defects which are inevitably present in integrated circuit wafers. It is these defects which currently limit chips sizes to around 0.5 inches on a side, since larger chips generally have uneconomic yields. The yield of a WSI circuit would almost certainly be zero if no precautions are taken to tolerate the defects.

Defects in integrated circuits can be classified into three groups according to Peltzer (1983). These are *gross, intermediate* and *random* defects. Gross defects affect large portions of the wafer and are caused by problems during processing such as poor lithographic alignment, incorrect doping level and missed process steps, etc. Intermediate defects are more localised, typically a few millimetres in size, and are caused by dust particles, and scratches, etc. Random defects are often called *point* defects and are caused mainly by pinholes in oxide layers, lithographic defects, and defects in the crystal lattice of the silicon. Random defects are the most important class in the context of WSI. Gross defects cannot be tolerated by any fault tolerant scheme, while intermediate defects are best dealt with by improvements in cleanliness and handling during processing.

Typically, random defects occur with a density between 2 and 5 defects per square centimetre. However, the distribution of defects over the wafer is not entirely random, and the defects tend to form clusters (Stapper 1986). This means that there are regions having a much higher than average defect density and other regions having relatively few defects. It is chips occurring in the latter regions which produce the functional devices. In WSI the clustered defects must be tolerated if useful device yields are to be achieved.

Another problem of WSI is that even with suitable configuration techniques, there is still some circuitry which cannot be protected easily from defects. This includes most global signal lines such as clocks, switch control lines and power supplies. It may also be necessary to disconnect a processor from its power supply if a fault causes a short circuit between power rails within the processor. These problems are considered by Warren *et al.* (1986).

16.4 Classification of hardware fault tolerance schemes

Nearly all the work on WSI has focussed on arrays of identical processing elements, both linear and two-dimensional. The reason for this is that in such arrays, a pool of spare processing elements can be incorporated into the array and used to replace those elements which are defective. In this chapter we consider these techniques and place particular emphasis on the fabrication of two-dimensional arrays.

Each WSI implementation technique can be classified in two ways:

1. according to the strategy for fault avoidance defined by the way in which the switches are organised
2. according to the way in which switches are implemented

Any particular approach to WSI will therefore be a member of a class from each category. The classes of switch organisation scheme are shown in Figure 16.1 while the methods of switch implementation are shown classified in Figure 16.2.

The switch implementation classification is essentially that of Katevenis and Blatt (1985) and is presented from left to right in order of increasing *lateness of*

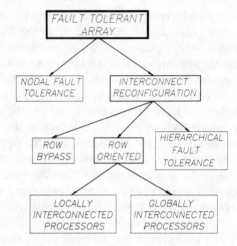

Fig. 16.1 *Classification of switching schemes for fault-tolerant strategies in two-dimensional processor arrays.*

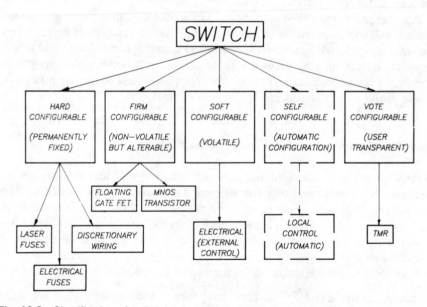

Fig. 16.2 *Classification of switch implementation methods used in WSI circuits.*

binding. This means that those techniques on the left hand side of the tree are fixed at the time of manufacture or configuration and are essentially permanent for the rest of the life of the device. Switch implementations further to the right become fixed progressively later in their life and have increasing facilities for re-configuration.

16.5 Switch organisation and configuration schemes

In this section we briefly describe some of the approaches to the configuration of arrays of processing elements which fall into the classes shown in Figure 16.2.

16.5.1 Nodal fault tolerance

The objective of Nodal fault tolerance is to increase the yield of the individual processing nodes (or elements) within an array to almost 100% so that an acceptable overall array yield is achieved without having to configure the connections between nodes. The most straightforward method for doing this is to use triple modular redundancy (TMR). TMR involves using three processors in place of each of the original single processing elements, together with a voting circuit. The voting circuit provides the output of the TMR node by delivering the majority verdict of the outputs of the three processors, all of which execute the same function on identical input data. In this way any one of the processors

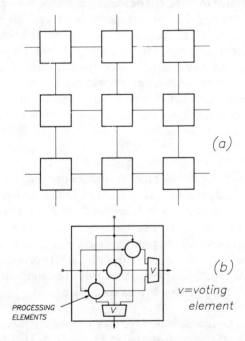

Fig. 16.3 *Technique of triple modular redundancy: (a) TMR array, (b) TMR node.*

can be faulty, but the voting circuit will still ouput the correct result. A TMR fault tolerant array is shown in Figure 16.3(a) with an individual TMR node being shown in Figure 16.3(b).

The TMR scheme has the advantage that it is very simple to implement since no configuration of the processors in the array is carried out and no testing of the processors is required. It also has the advantage that it can tolerate in-service faults, both permanent and intermittent, without the user being affected or even needing to be aware that a fault exists. However, against these advantages there are some serious drawbacks. TMR systems require a large hardware overhead since each processor is now triplicated. Furthermore, the statistics involved in 2-out-of-3 majority voting schemes indicate that the yield of the processors involved must be quite high in order to gain any nodal yield advantage at all from using the scheme, and that even at best, the gain in yield is not dramatic. For example, a processor yield of at least 50% is required before the node yield exceeds the processor yield. Furthermore, the maximum gain in yield is achieved when the original processor yield is about 87%, at which point the node yield has risen to 95%.

The very high hardware overhead of TMR for a small gain in yield has prompted the study of other methods of implementing nodal fault tolerance. One alternative method is to use two processors in place of the original processor, as against three for TMR. The idea is then to use a switch to select only one of the two processing elements for use in the array. This approach requires that the user knows which of the processing elements is working correctly and therefore implies that testing of the processing elements must be carried out. This could be done either by external test or by some form of self-test procedure. A technique using two processors per site was successfully used by Grinberg *et al.* (1984) to increase the yield of individual wafers in their three-dimensional computer based on stacked wafers. They used discretionary wiring to select between the two PEs on the node.

16.5.2 Row or column replacement schemes

As we have seen, nodal fault tolerance minimises or even eliminates the need to consider how to configure an array to avoid faults, and tries to sufficiently increase the yield of the nodes so that they can be used directly in an array. Most hardware fault tolerance techniques, however, rely upon some form of alteration to the connections between processors in order to generate a subset of the main array which is fully functional. In this way, faulty processors are completely isolated from the functional part of the array. The simplest technique of this type is the row-selection or row-bypass method.

Row selection: The row selection technique is widely used in memory chips for yield enhancement and is illustrated in Figure 16.4. There are many ways in which the row selection procedure can be organised and several of these are discussed by Fitzgerald and Thoma (1980). The idea is to incorporate spare rows (or columns) into the array of memory elements and to use these rows to replace any rows from the main array which are found to contain faults. This technique is very simple to implement in memory chips because there are no signal interconnection paths between cells, and spare rows can be selected simply by programming the decoding circuitry appropriately. The decoder is often pro-grammed by blowing electrical fuses.

Many memory manufacturers claim that the row selection technique is useful in the early stages of production of a device for increasing yield and give figures ranging from a 30-fold yield increase in immature processes, reducing to a 1.5-fold yield increase in a mature process (Smith, 1981). However, NEC claim not to need fault tolerance at all (see Posa, 1981 and Rogers, 1982.).

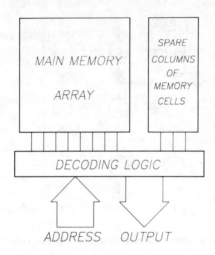

Fig. 16.4 *Schematic of fault-tolerant memory showing technique of row de-selection.*

Row bypass: In memory chips, sufficient rows are simply selected from those available so that sufficient cells are available for storing information. In pro-cessor arrays, a similar technique can be applied, but in addition, the connec-tivity between processing elements must be maintained. This means that when a row containing a fault is de-selected, its input signals must also be diverted to an alternative, functional row. This can be achieved most simply by employing bypass circuitry around each row so that the whole row can be bypassed in the event of it containing one or more faults. All spare rows are initially bypassed and the bypass is removed when the row is brought into operation.

A row bypass scheme like that described above was proposed by McCanny and McWhirter (1983). Figure 16.5 illustrates their approach and shows how multiplexers are used for the bypass mechanism.

Moore *et al.* (1986) extend the basic row bypass technique by considering the effect of an imperfect multiplexer yield on the overall array yield. They propose various modified bypass circuits, some involving more than one multiplexer per cell, which are able to tolerate many of the faults which could occur in the bypass circuitry.

An important requirement of the bypass technique is that the yield of the individual processing elements is very high so that there is only a small number of faults in the array compared with the number of rows in the array. It is this constraint which enables the simple method of discarding entire rows to be beneficially employed. If too many faults are present in the array, it is likely that many or even all of the rows will contain faults, and no increase in yield will be achieved.

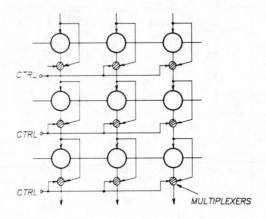

Fig. 16.5 *Row bypass scheme of McCanny and McWhirter (1983).*

16.5.3 Hierarchical fault tolerance
Hierarchical fault tolerance techniques have similarities to nodal fault tolerance. In the scheme proposed by Hedlund and Snyder (1982), illustrated in Figure 16.6, processing elements are grouped into blocks of twelve, out of which only four are required to work. The four working processors are then interconnected as a 2 × 2 subarray within each block and the blocks are then interconnected to form a two-dimensional array. If any block in a row of blocks is unable to configure a 2 × 2 subarray, the whole row of blocks is bypassed. This scheme offers two levels of hierarchy and potentially allows a functional array to be configured from an array containing a large number of faulty devices, but it also requires a very large overhead of redundant processors. It should be noted that although Hedlund and Snyder have chosen to bypass a whole row of blocks if a single block in that row cannot be configured into a 2 × 2 subarray, it would also be possible to use a more sophisticated strategy, such as one of those described in the next sections, for avoiding faulty blocks. In this way an improved yield characteristic might be achieved.

16.5.4 Row generation schemes
In these fault-tolerant schemes, the idea is to generate functional rows of processing elements in which each functional row is constructed by taking one

functional processor from each column of the array. The functional rows are then interconnected down the columns in the vertical direction to form the two-dimensional array. Any faulty or unused processors encountered when interconnecting down a column are bypassed. Several row generation schemes are presented in the literature by Sami and Stefanelli (1983), Moore and Mahat

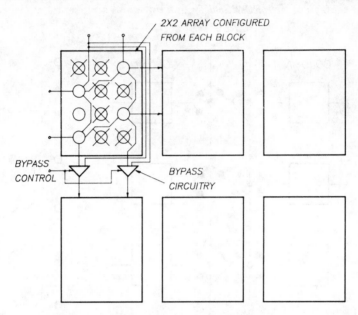

Fig. 16.6 *Hierarchical scheme of Hedlund and Snyder (1982).*

(1985), Evans (1985) and Bentley and Jesshope (1986). One of the schemes of Moore and Mahat is reproduced in Figure 16.7 together with two columns of processors which have been configured by the techniques.

The scheme of Sami and Stefanelli is similar to that of Moore and Mahat but allows as many row shifts as necessary. It therefore has superior performance but due to the high overhead is most suitable for arrays with a small number of spare rows. A detailed description of the scheme of Evans (1985) is presented in Chapter 17.

16.5.5 Global organisation
Schemes classified under the heading of global organisation offer a much greater flexibility as to how the cells are interconnected than other approaches. the most general scheme is probably that proposed by Katevinis and Blatt (1985), which is reproduced in Figure 16.8.

The idea of global organisation is to provide the array with buses which run the entire length of the array between both the rows and columns. Switching nodes are inserted at the intersections of the buses so that connections can be selectively made between separate buses, and between buses and processing elements. In principle the global nature of the scheme means that a processor anywhere in the array could be connected to any other processor. This would

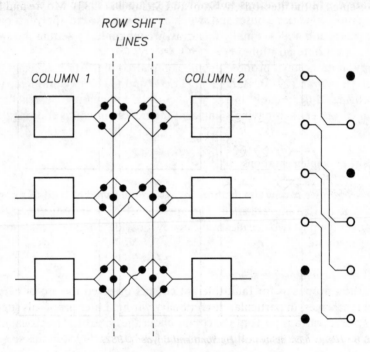

Fig. 16.7   *Row-oriented configuration scheme of Moore and Mahat (1985).*

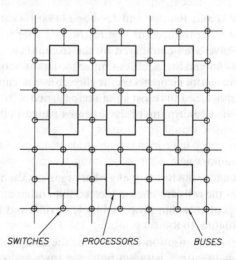

Fig. 16.8   *Global interconnection scheme of Katevenis and Blatt (1985).*

provide an excellent ability for avoiding faults, but could lead to timing problems due to extra transmission delays being introduced into signal lines. The operation of a global configuration scheme can be described generally as follows. First, the buses are tested by an external tester, and then the working buses are used to give access to the switching nodes. These are tested and the combination of working buses and switching nodes is used to apply test patterns to the processing nodes. Finally the array is configured by setting the switches to the appropriate positions.

Many authors have proposed similar global configuration schemes. These include Hsia *et al.* (1979), Raffel *et al.* (1983) and Gaverick and Pierce (1983). The schemes differ mainly in the way which the switching elements are implemented rather than having significant differences in configuration strategy.

16.6 Switch implementations

In this section we review the methods by which the switches used in a configuring scheme can be implemented. The approaches have improved very significantly over the past two decades and can now provide a highly reliable interconnection medium.

16.6.1 Hard-configurable schemes

The earliest proposals for fault-tolerant circuits involved the use of hard configurable schemes, in particular discretionary wiring. Later proposals suggested using fuses at various parts of the circuit under the control of electrical heating or laser cutting. The laser cutting technique has been included in this section on hard configurable switching schemes since it has until recently been an irreversible process. However, recent reports indicate that the laser cutting processes may be reliably reversed, and this process will also be discussed.

Discretionary wiring approaches: The earliest attempts at increasing the area of integrated circuits were based on the principle of discretionary wiring. The idea is to place more circuit elements on the chip or wafer than actually required to perform the function and to test each of these elements by probing the wafer. The results of the test can then be represented as a wafer map and a metal mask can be designed which would interconnect the working devices. Sack *et al.* (1964) propose this approach for enabling whole wafer circuits to be produced. They demonstrate a complete wafer containing 108 gates interconnected in the form of a shift register. There are many variations on the basic discretionary wiring technique. Some are described in Petritz (1967), Lathrop *et al.* (1967) and Calhoun (1969). All of these approaches appear to offer advantages at the level of integration available at the time (about 5000 gates on a 1 inch diameter wafer).

One of the problems with discretionary wiring is that it relies on there being a very few faults in the wiring layer, which although achievable at the device geometries of the late 1960s, is unlikely to be successful at 1 μm geometries and with 4-6 inch diameter wafers. Another problem is that the probe testing of the devices on the wafer causes damage to the wafer surface which increases the probability of a fault occurring during subsequent processing. An interesting approach along the theme of discretionary wiring is the one proposed by Barsuhn (1978), in which he fabricates a wafer of memory chips. Faulty chips are replaced by good, individual mirror image chips which are flip-chip bonded over the faulty device. Barsuhn claims success with this method for a 2.25 inch diameter wafer.

Electrical fuses: Electrical fuses are commonly employed as the method of implementing the necessary switching in yield enhancement techniques for memory chips. They have also been extensively used in PROMs and programmable array logic (PALs) for defining logic functions, although erasable techniques based on stored charge have now largely taken over. The electrical fuse technique is based upon heating the fuse, which is commonly made of aluminium or polysilicon so that it melts and creates an open circuit. When used in conjunction with pull-up or pull-down components, a change in logic level can be achieved and subsequently used to control other circuits. Although such fuses can in principle be combined in a circuit to allow a reversal of the effect of blowing a fuse by blowing a second, the fusing process itself is essentially irreversible. This means that the fusing technique cannot be used to isolate parts of a wafer for testing purposes and subsequently reconnect them.

Laser linking and cutting: The technique of using a laser beam to either cut or weld signal paths has been extensively studied at MIT Lincon Laboratory by Raffel and his research team in the Restructurable VLSI (RVLSI) approach to large area integration (Raffel *et al.*, 1983). The approach now seems to be a strong contender for WSI due to its reliability and ease of execution. The structures used for linking and welding, together with details of the procedures used and some results are presented by Chapman (1985).

The MIT structure for making links between first and second metal layers is reproduced in Figure 16.9 and details of the laser pulse required and the link parameters are given in Table 16.1. During the linking process, for which an argon laser focussed to a 10 μm spot size is used, successive melting of the second layer metal, the amorphous silicon insulator and part of the first layer metal occurs. This creates a silicon-aluminium alloy which provides the conducting path. An important feature of the melting process is that it occurs over a relatively long time period with a low power pulse (1 ms as against 100 ns for commercial laser cutting systems). This avoids the splatter which normally accompanies metal vaporisation. MIT claim that the failure rate of the process is below that of the processing defects occurring during link fabrication.

The link structure described above also enables cuts to be made by using the laser to melt either the first or second layer metal just before it enters the link structure. Cuts have been successfully carried out using the same low power as that used for linking so that splatter is avoided.

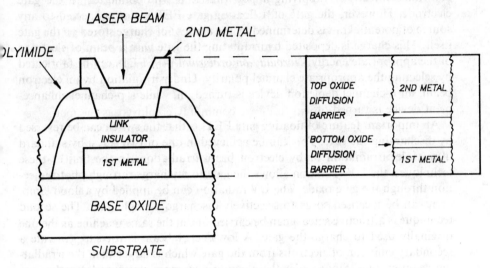

Fig. 16.9 *Cross-section of laser link from Chapman (1985).*

Table 16.1: *Laser link parameters*

Laser Power	$> 1 \cdot 2\,\mathrm{W}$
Pulse Width	$\approx 1\,ms$
Open Link Resistance	$> 10^{14}\,\Omega$
Formed Link Resistance	$< 1\,\Omega$
Failure Rate	$< 0 \cdot 01\%$
Capacitance	$\approx 35\,f\mathrm{F}$

16.6.2 Firm-configurable switching schemes

These schemes are characterised by switch reversibility combined with non-volatility of the switch setting. From the point of view of testing and configuring a wafer, this type of switching scheme is attractive, since areas of the wafer can be temporarily isolated while a detailed local test is carried out. Mistakes in configuration, or faults occurring after configuration can also be conveniently dealt with. There are two main approaches to firm configurable switches: the Floating Gate FET and the MNOS transistor. These are described more fully in the following paragraphs.

Floating gate FET: The operation of a floating gate FET switch is similar to that of a normal FET in that it is the voltage on the gate of the FET which determines whether the transistor is on or off. The difference is in the manner in which the gate voltage is applied. In a normal FET the gate voltage is controlled directly by applying a potential to a wire connected to the gate electrode. However, the gate of a floating gate FET is not connected to any source of potential but is determined by the amount of charge stored on the gate itself. This charge is deposited by irradiating the gate with a beam of electrons of the appropriate energy. *Normally-on* or *normally-off* FETs can be fabricated by selecting the appropriate channel polarity. Under irradiation by an electron beam, an n-channel depletion device is turned on, while a p-channel enhancement device is turned off.

An important feature of floating gate FETs is that the switch can be reversed by discharging the gate. This can be achieved in one of two ways: by standard ultra-violet irradiation or by electron beam irradiation. In the first of these techniques, the UV radiation allows the gate to discharge through photo-injection through the gate oxide. The UV radiation can be applied by a flood lamp, or it can be localised so as to selectively discharge a single gate. The second technique is attractive since it can be carried out in the same machine as the one originally used to charge the gate. A low-energy beam is used to generate a secondary emission of electrons from the gate which is larger than the irradiating beam current. Since under these conditions, more electrons leave the gate than arrive at it, the charge on the gate reduces.

Although the floating gate FET switch is very attractive and will probably be acceptable for commercial devices, the retention time of charge on the gate may be too short for military devices (Shaver, 1984). However, the use of the floating gate FET in wafer scale integration is being investigated as part of the ESPRIT project number 824. An overview of this project is presented by Trilhe (1987).

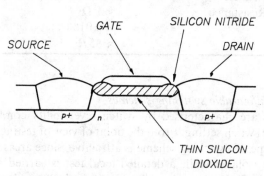

Fig. 16.10   *Structure of MNOS transistor.*

The MNOS transistor switch: The MNOS transistor illustrated in Figure 16.10 is commonly used in Electrically Alterable Read Only Memories (EAROMS). These devices can store information for many years but can also be altered in a simple manner by the application of the appropriate programming signals which tend to be about 25 to 40 volts. The programming voltages cause injection of electrons into the boundary region between the silicon nitride layer and the silicon dioxide layer. When the programming voltage is removed, the charge is retained since the boundary region is isolated. Erasure is achieved in a similar manner, with stored charge being repelled from the boundary and absorbed into the substrate. (see Muroga, 1982)

Although the MNOS transistor switch is simple in operation, it does have drawbacks in the context of wafer scale integration. The main problem is that a connection to control the programming would be required for each transistor and these would have to be accessible from the edge of the wafer. For small numbers of switches this may be feasible, but for large numbers, the problem will be serious.

16.6.3 Soft-configurable switching schemes
The main type of soft-configurable switching scheme uses externally controlled electrical switching elements. This type of switch implementation is probably the one which most people would first think of. The idea is to design the switching nodes using ordinary logic gates. These are then controlled from an external source so that the desired connections are made between the processors. The great advantage of electrical switches is that they use only standard circuit components which are the same as those used for the remainder of the circuitry. In addition, since no specialised equipment is required, configuration can potentially be carried out in the field if in-service faults occur. However, their main disadvantage is the same as for MNOS transistor switches, that the wiring needed to control them can become a serious problem. This type of switch implementation technique has, however, been used successfully by Anamartic (formerly Sinclair Research) in their wafer scale disk memory. Their switches are configured under the control of an external tester and the scheme is described in Aubusson and Catt (1978).

16.6.4 Vote-configurable switching schemes
Circuits of this type have already been considered under the heading of nodal fault tolerance. However, although it is not immediately obvious, it is worth remembering that they are a form of electrical switch. Their advantage over externally controlled switches is that no wiring or global control is required; all the information they require to output the correct result is available locally. It is unfortunate that the hardware redundancy associated with the processing node to be used with this type of switch is so high as to make it impracticable in most cases, especially where the yield of the individual processors is low.

16.6.5 Self-organising switching schemes
Self-organising switching schemes combine the convenience of implementation offered by externally controlled switches with potential for reconfiguration in the field and have the further advantage that no external control of the switches is required. This not only removes the need for large numbers of extra pins just

for configuration purposes, but also means that external computation to cal-culate the desired configuration pattern is not necessary since the ability to make descisions about the configuration pattern resides within the array cells them-selves.

Research carried out by the author at the Royal Signals and Radar Establish-ment, Malvern, UK has resulted in the development of a self-organising scheme for two-dimensional, orthogonally interconnected processor arrays called WIN-NER, an acronym for wafer integration by nearest-neighbour electrical recon-figuration (Evans, 1985). Each processor in the array is provided with a small amount of control logic, equivalent to about 20 gates. This extra logic allows the processing elements to communicate with each other and come to a collective decision about how the working processors should be interconnected to form a functional two-dimensional array. The configuration of the array is achieved without any exernal assistance and is guaranteed to produce a stable configur-ation without any contention problems between processors. The WINNER technique is described in detail in the following chapter.

16.7 WSI demonstrators

Much of the research which has been carried out on wafer scale integration has been limited to paper exercises backed up in many cases by computer simu-lations. There have been relatively few examples of actual devices being built, although this is now changing and several demonstration devices are currently being developed. In this section we describe some of the devices which have been fabricated.

16.7.1 Trilogy
Trilogy is probably the best known company involved in WSI. Their very ambitious project to build an IBM-compatible mainframe computer on a single wafer received much attention in the press. The circuit was partitioned into about 1 500 blocks containing between 10 and 50 gates each and triple modular redundancy was the method used to increase the block yield. In order to reduce the effect of clustered defects, the triplicated blocks were not placed adjacent to each other.

Unfortunately, Trilogy were unsuccessful in their attempt and investors have been wary of WSI ever since. According to Peltzer (1983), two main problems can be identified. The first is that the spacing of the triplicated blocks resulted in an increased transmission time between blocks and the goal of an IBM-compatible device was not achieved due to lack of speed. The second is that the technology chosen for implementing the wafer was ECL, and resulted in a power consumption of about 1 kW on a 4 inch wafer. This led to serious problems of thermal management.

16.7.2 Anamartic and the solid state disk memory

In the UK the wafer-scale memory built by Anamartic is probably the best known. The reason for this is that right from the start the device has been specifically aimed at the consumer market and as a result has received much attention in both the technical and national press. In addition, a novel configuration technique has been used and this has captured the attention of many people. Working wafers with 0.5 M Bits of storage were demonstrated at the Southampton Workshop on Wafer Scale Integration in 1985. A higher-density wafer is currently being developed as a commercial product which Anamartic hope will be able to replace conventional disks and have both much improved reliability and access time.

The technique employed by Anamartic for enabling the wafer to provide sufficient yield is commonly called the *Catt spiral* which was proposed by Aubusson and Catt (1978). The scheme generates a linear array of interconnected memory blocks starting from one block at the edge of the wafer and adding extra blocks to the chain one by one. The configuaration is implemented using electrical switches which are controlled by an external tester/controller.

The implementation procedure operates as follows. The controller initially tests one of the chips on the periphery of the array. If the chip is faulty, another peripheral chip is chosen until a functional device is found. Instructions are then sent to the functional chip to tell it to connect itself to one of its neighbouring chips. (The way in which the neighbour is chosen is described later.) The chosen neighbour is then tested by the external controller by sending test patterns through the first chip, and into the neighbour. Test results are passed along the reverse route. If the neighbour is faulty, an alternative neighbour is chosen until a functional neighbour is found. This chip is then added to the chain. The chain is then further built up by repeating the process. If at any point in the configuration procedure a chip at the end of the chain is found to have no functional neighbours, it is removed from the chain and the previous chip in the chain is used as the new chain end. This backtracking ability can also enable the chain to escape from dead ends which may exist on the array.

The order in which neighbours are selected as candidates for the next position in the chain determines the shape of the final chain. In the Anamartic design, the most right-hand neighbour is selected, and this results in a chain of good chips which hugs the outer edge of the array and spirals in towards the centre of the wafer. Figure 16.11 shows a wafer which has been configured in this way. For wafer scale integration the approach has the advantage that all the interconnects between elements of the chain are checked at the time of testing, and as a result a chip cannot be included in the chain unless all wires are intact. Another advantage of the way in which the linear array is built up block by block is that the external switch control signals are applied serially. This avoids the need for large numbers of pins on the wafer.

Working prototype devices have been successfully fabricated and have demonstrated that the yield of the control circuitry on the cells is adequate, with around 90% of cells having working circuitry. Few details of performance have been published, but in a public demonstration of a device, it was clear that a large proportion of the memory elements were also functional and could be connected to the spiral chain.

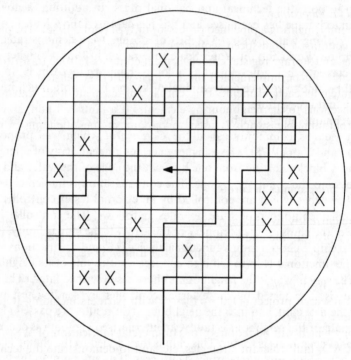

Fig. 16.11   *Wafer configured as a linear processor array using the spiral technique of Aubusson and Catt (1978).*

The use of electrical switches means of course that the spiral pattern generated by the test and configuration procedure is volatile and will need to be reapplied each time the device is powered up. The device could be retested each time, but alternatively the configuration pattern can be stored in a ROM. The contents of the ROM can then be loaded into the wafer before it is used.

16.7.3 MIT and Lincoln Laboratory
The work at Lincoln Labs on wafer scale integration is based on their technique of restructurable very large scale integration (RVLSI). Several demonstrators have been built successfully and a review of the progress of the project is presented in Rhodes (1986).

The first demonstrator based on the RVLSI approach was a digital integrator consisting of 256 10-bit counters partitioned into 64 cells. Each cell contains four 10-bit counters. The complete device contains 130 000 transistors on a 4 inch wafer using 5 μm CMOS technology. The configuration switches use 1 900 laser anti-fuses and 137 laser fuses.

More recently, a Multiply-Accumulative device has been fabricated.

16.7.4 GTE laboratory

GTE are implementing a pipelined processor in WSI (Cole, 1985). The pipelined element contains a high-speed sequencer, a micro-code RAM, a 32-bit ALU, and status and storage registers. Each element contains 150 000 transistors, and 60 elements can be implemented on a 3 inch wafer. A self-test procedure is incorporated in each element. This checks the element itself and also the interconnections to the neighbouring elements. If a fault is found, an electron beam programmable switch is used to disconnect the offending processing element.

16.8 WSI programmes in europe

There are currently two major European research projects on WSI, within the UK Alvey programme and within the European Esprit programme. The aims of these projects are described briefly in the following sections.

16.8.1 The Alvey project on WSI

This UK research project is led by Plessey with partners including ICL, GEC and Brunel University. The aim of the project is to develop the enabling technology which will be required for commercial fabrication of WSI devices, such as CAD, fault-tolerant strategies, etc, and to demonstrate the technology by fabrication of two demonstrator devices.

The main candidates for the demonstrators are the WASP (Word Associative String Processor) architecture being developed by Brunel University, (Lea, 1985), and a fine-grain systolic array architecture from GEC. The WASP architecture is essentially a linear array which has been mapped elegantly on to the wafer surface, while the systolic demonstrator involves a two-dimensional array of processing elements.

16.8.2 The Esprit project on WSI

The Esprit project on WSI is a collaboration between six European organis-ations led by Thomson Semiconductors of France and including British Tele-com of the UK. A review of the project is presented by Trilhe (1987).

Three demonstration devices are planned:

- a 4.5-M Bit static RAM
- a programmable systolic array
- a fault-tolerant microprocessor

All three devices have been designed not simply to demonstrate the WSI concept but also to be used in systems if successful. The fault tolerant microprocessor is an interesting demonstrator in that it represents one of the few attempts to introduce fault tolerance into a non-regular structure.

16.9 Conclusions

In this necessarily brief introduction to WSI we have described the motivation behind the work and some of the techniques published in the literature for configuring arrays of processors.

The existence of two major European projects on WSI, both with 50% funding from the companies involved is encouraging and indicates that there is a will to succeed.

References

AUBUSSON, R. C. and CATT, I. (1978): Wafer scale Integration – a Fault Tolerant Procedure, *IEEE J. Solid State Circuits*, **13**, (No. 3), pp. 339–44

BARSUHN, H. (1977): Functional Wafer – a new step in LSI, *Proc. European Solid State Circuits Conf.* Sept 1977, pp. 79–80

BENTLEY, L, and JESSHOPE C. R. (1986): The implementation of a Two-dimensional Redundancy Scheme in a Wafer Scale, High Speed Disk Memory, in JESSHOPE, C. R. and MOORE, W. R. (eds.) *Wafer Scale Integration*, (Adam Hilger, Bristol), pp. 187–197

CALHOUN, D. F. (1969): The Pad Relocation Technique for Interconnecting LSI arrays of imperfect Yield, *Proc. Fall Joint Computer Conf.* pp. 99–109

CHAPMAN, G. H. (1985): Laser-Linking Technology for RVLSI, *Proc. Int. Workshop on Wafer Scale Integration*, Southampton University, July 1985, pp. 204–15

COLE, B. C. (1985): Wafer Scale Integration Faces Pessimism, *Electronics Week*, (April), pp. 49–53

EVANS, R. A. (1985): A self-Organising Fault-Tolerant 2-dimensional Array, *Proc. VLSI–85*, Tokyo, 1985, pp. 239–48

EVANS, R. A. and McWhirter, J. G. (1986): A Hierarchical Test Strategy for Self-organising Fault-tolerant Arrays, in MOORE, W. R, MCCABEA, P. H. and URAUHART, X. X. (eds.) *Systolic Arrays*, (Adam Hilger), Bristol, pp. 229–38

FITZGERALD, B. F. and THOMA, E. P. (1980): Circuit Implementation of Fusible Redundant Addresses in RAMs for Productivity Enhancement, *IBM. Res. Develop.* **24**, (No. 3), pp. 291–8

GAVERICK, S. L. and PIERCE, E. A. (1983): A single Wafer 16-point 16 MHz FFT Processor, *Proc. 1983 Custom Integrated Circuits Conf.*, IEEE, pp. 244–8

GRINBERG, J, NUDD, G. R. and ETCHELLS, R. D. (1984): 3D Computing Structures for high throughput information processing, in CAPPELLO, P. R. (ed.), *VLSI Signal Processing*, (IEEE Press, NY), pp. 2–14

HEDLUND, K. S. and Snyder, L. (1982): Wafer Scale Integration of Configurable Highly Parallel (CHiP) Processor, *Proc. 1982 Int. Conf. on Parallel Processing*, IEEE, pp. 262–4

HSIA, Y, CHANG, G. C. C. and ERWIN, F. D. (1979): Adaptive Wafer Scale Integration, *Proc. 11th Conf. on Solid State Devices*, Tokyo, Japan. *J. Appl. Phys.* **19**, pp. 193–202, (Supplement 19-1)

KATEVENIS, M. G. H. and BLATT, M. G. (1985): Switch Design for Soft-Configurable WSI systems, *Proc. 1985 VLSI Conf.* pp. 197–219

LATHROP, J. W. CLARK, R. S., HULL, J. E. and JENNINGS, R. M. (1967): A Discretionary

Wiring syustem as the Interface between Design Automation and Semiconductor array Automation, *Proc IEEE*, **55**, pp. 1988-7

LEA, R. M. (1985): WASP – A WSI Associative String Processor for Structured data processing, in JESSHOPE, C. R. and MOORE, W. R. (eds.) *Wafer Scale Integration*, (Adam Hilger, Bristol), pp. 140-7

McCANNY, J. V. and McWHIRTER, J. G. (1983): Yield Enhancement of Bit level Systolic Array chips using Fault Tolerant Techniques, *Electron. Lett.*, **19**, (No. 14), pp. 525-7

MOORE, W. R., McCABE, A. P. H. and BAWA, V. (1986): Fault Tolerance in a Large Bit-level Systolic Array, in JESSHOPE, C. R. and MOORE, W. R. (eds.), *Wafer Scale Integration*, (Adam Hilger, Bristol), pp. 259-72

MOORE, W. R. and MAHAT, R. (1985): Fault Tolerant Communications for Wafer Scale Integration of a Processor Array, *Microelectronics and Reliability*, **25**, (No. 2), pp. 291-4

MUROGA, S. (1982): *VLSI Systems Design*, (John Wiley and Sons, New York), pp. 264-6

PELTZER, D. (1983): Wafer Scale Integration: The Limits of VLSI?, *VLSI Design*, (Sept.) pp. 43-7

PETRITZ, R. I. (1967): Current Status of LSI Technology, *IEEE J. Solid State Circuits*, **SC-2**, (No. 4), pp. 130-47

POSA, J. G. (1981): What to do when the Bits go out, *Electronics*, (28 July), pp. 117-20

RAFFEL, J. I, ANDERSON, A. H., CHAPMAN, G. H., GAVERICK, S. L., KOUBLE, K. H., MATHUR, B. and SOARES, A. M. (1983): A Demonstration of very large area Integration using Laser restructuring, *Proc. 1982 Int. Symp. on Circuits and Systems*, IEEE, pp. 781-4

RHODES, F. M. (1986): Performance Characteristics of the RVLSI Technology, *Proc. IFIP Workshop on Wafer Scale Integration*, Grenoble, France, pp. 31-42

ROGERS, T. J. (1982): Redundancy in RAMs, *Proc. Int. Solid State Circuits Conf.*, February 1982, pp. 228-9

SACK, E. A. LYMAN, R. C. and CHANG, C. Y. (1964): Evolution of the Concept of a Computer on a slice, *Proc. IEEE*, **52**, (Dec. 1964), pp. 1713-20

SAMI, M. and STEFANELLI, R. (1983): Reconfigurable Architectures for VLSI Processing Arrays, *AFIPS 1983) National Computer Conf.* Anaheim, California, pp. 565-77

SHAVER, D. C. (1984): *Solid State Technology*, pp. 135-9, (Feb.)

SMITH, R. T. (1981): Using a laser beam to substitute good cells for bad, *Electronics*, (28 July), pp. 131-4

STAPPER, C. H. (1986): On yield, Fault Distributions and Clustering Particles, *IBM J. Res. Develop.* 30, (No. 3), pp. 326-8

TRILHE, J. (1987): An European program on Wafer Scale Integration, *Proc. VLSI—87*, Vancouver, (Aug.) pp. 293-308

WARREN, K. *et al.* (1986): Power, Clock and Signal distribution in the WASP Device, *Proc. WSI Workshop*, Grenoble, (Mar.)

WSI design techniques and applications of WSI

Chapter 17

Self-organising arrays for wafer scale integration

R. A. Evans

17.1 Introduction

In the review of configuration techniques for wafer scale integration (WSI) presented earlier we have seen that many of the approaches to hardware fault tolerance in two-dimensional arrays involve the use of electrical switching networks to allow faulty processors to be replaced by spares. In all cases the switches are set by some external controller responsible for deciding which switches should be used and how the array should be configured. At the Royal Signals and Radar Establishment, Malvern, UK, we have been investigating techniques which enable an array to automatically configure itself around the faulty elements and generate a functional two-dimensional array from an array containing many faults. Such a scheme would avoid the need for an external controller and would also provide a system which could readily be reconfigured if a fault occurs in service.

We have developed a novel solution to the two-dimenstional array configuration problem. In our technique, which we call WINNER* (Evans, 1985), each cell within the array is provided with some intelligence in the form of a small amount of additional control circuitry. This is sufficient to enable each processing element to independently and simultaneously make local decisions about how it should be connected to neighbouring elements, taking into account its own functionality, the functionality of its neighbours, and the connection priorities to these neighbours which are defined in specific algorithms. The effects of these local decisions propagate throughout the array and manifest themselves globally as a complete self-organisation of the functional processing elements into a correctly interconnected functional two-dimensional processor array. We call such arrays *self-organising arrays.*

A number of algorithms incorporating the concepts of self-organisation can be derived. For the purposes of this book we concentrate our attention on two related algorithms which illustrate the technique. The first scheme applies the WINNER algorithm in one dimension of the array only, and generates func-

---

* WINNER is an acronym for Wafer Integration by Nearest-Neighbour Electrical Reconfiguration.

tional rows of processors. A simple fault bypassing technique is then used to form the second dimension (Evans, 1985). It is the simpler of the two algorithms and is presented in 17.3. The second method applies the WINNER algorithm in two dimensions and is discussed in 17.4, (Evans, *et al.*, 1985).

We first describe the algorithms at a high logical and operational level. Detailed circuit level descriptions and results of performance simulations etc., are presented later.

17.2 Definitions

The following terms will be used in subsequent description of the algorithms:

Processor: This is the circuit which performs the node function when the array is operating in-service. It communicates with four neighbours to the north, south, east and west. (The assumption that the processor has only these or-thogonal connections does not limit the generality of the approach since all other nearest-neighbour array interconnection structures can be reduced to the orthogonal form. An example of a hexagonally interconnected array and its orthogonal equivalent are shown in Figure 17.1).

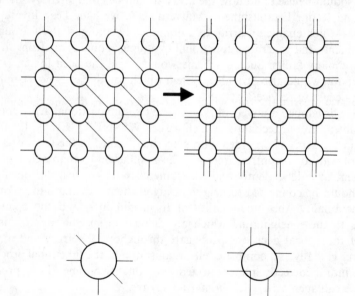

Fig. 17.1 *Hexagonally interconnected array and its orthogonal equivalent.*

Control circuit: This comprises the extra logic which is added to the processor to provide it with the self-organisational ability.

Cell: A cell is the combination of the processor and its control circuitry and is illustrated schematically in Figure 17.2.

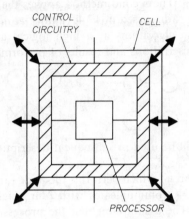

Fig. 17.2 *Relationship between cell, control circuitry and processor.*

17.3 Algorithm 1: WINNER in one dimension

The WINNER algorithm in one dimension is so called because it results in an array in which a number of functional *rows* have been generated. These rows avoid faulty cells but do not themselves form a two-dimensional array. The second dimension of the array is formed by interconnecting the functional rows in the vertical direction, bypassing any faulty cells. In order to describe the algorithm, we present an array which has been configured using the algorithm and explain how the configuration has been achieved by giving a simple pencil and paper description. We then show this procedure can be implemented as a circuit and describe the operation in detail.

Figure 17.3 shows the main stages of an array which is being configured by the one-dimensional WINNER algorithm. Figure 17.3(*a*) shows how an array with no faulty processors would be configured. Figure 17.3(*b*) shows an array containing faults which is to be configured in the example, with faulty processors being indicated by a cross. On paper the functional rows can be generated as follows, the cell numbers under consideration being referred to in brackets. Starting with the left-hand column of the array, choose the functional cell nearest to the top of the array (1). From this cell, look at the column to the right and make a connection to one of the three (in general) nearest-neighbour cells in that column, (in this case 2 or 7, since the upper boundary of the array limits the choice to two cells). In making the choice always choose a functional cell, with a preference for the cell nearest the top of the array. Repeat this procedure until the right hand side of the array is reached (2, 3, 4, 10). At this point, one complete functional row (1, 2, 3, 4, 10) has been generated. Subsequent rows are formed in a similar manner, treating the cells used in a previous functional row as if they were faulty cells (i.e. avoiding using them). The second row would therefore be (6, 12, 8, 9, 15). The procedure minimises the distance of the rows from the top of the array and therefore maximises the number of rows which can be generated on the array.

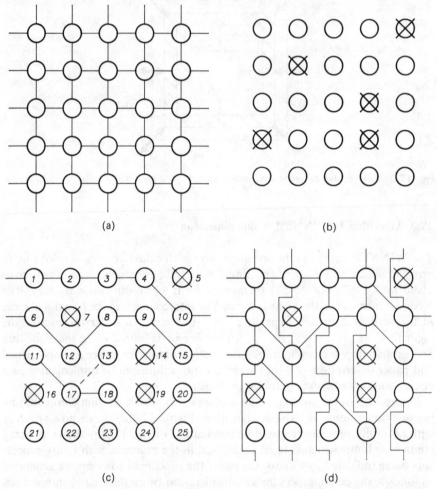

Fig. 17.3 *Configuration of a two-dimensional array using WINNER: (a) Configuration of a perfect array; (b) Array containing faults; (c) Rows configured around the faults; (d) Vertical connections made. Array configuration complete.*

In some rows a dead end may be encountered. This is illustrated in the generation of the third functional row in Figure 17.3(c) and occurs when processor 17 connects to 13. None of the neighbours of processor 13 are available for use and so the row must backtrack to 17 and try 18. The row can then be continued to completion. In a large array several back trackings may be required on some rows. A fourth row in this diagram cannot be constructed because a complete dead end is encountered. The row must therefore backtrack to the boundary of the array.

Each functional row formed in this way contains a processor taken from each column of the array. The second dimension of the required array can therefore be generated by making vertical connections between processors of each column and bypassing faulty processors and processors which although not faulty are nevertheless unused. This process completes the construction of the two-dimensional array and is shown in Figure 17.3(d). It is clear that the resulting array will be smaller than the original array due to the presence of the faulty elements but it should be noted that the x dimension of the functional array is identical to that of the given array, while the y dimension depends on the number of faults which occur in each column of the array.

17.3.1 Self organisation

We now consider how this pencil and paper procedure can be embodied within a circuit so that it can operate automatically. To simplify the problem we make the following assumptions:

1. Each processor contains some method enabling it to indicate reliably whether or not it is working. This could in principle be achieved by a self test procedure.
2. All connections between processors are fault-free
3. Each control circuit associated with a processor is fault–free
4. All connections between control circuits are fault–free
5. All data signals flow from left to right and top to bottom. This simplifies the description of the algorithms but in no way makes them less general

Items 2, 3 and 4 relate to the control circuitry being fault–free. As we shall see, the amount of circuitry required to implement the control logic is small and it should therefore have a high yield. However, research has been carried out on methods for tolerating control circuit faults and has shown that it is possible to tolerate all single faults and many multiple faults using a simple technique. An alternative approach (Evans and McWhirter, 1987) uses a modified scan-path testing technique to enable faulty control circuits to be identified by a simple externally applied test. Faulty control circuits can then be *masked* out of the array so that a correctly configured array can still be generated. Regrettably, space does not permit further discussion of these techniques in this book.

In order to enable faulty cells to be avoided and to allow control circuits in adjacent cells to communicate, a cell with greater connectivity than the original processor is required. A schematic view of a cell with sufficient connectivity to perform the one-dimensional self-organising function is shown in Figure 17.4.

It can be seen that although the north-to-south connection is unchanged, extra channels have been provided on the eastern and western sides for both inter-processor and inter-control-circuit communication. These connections allow the cell to communicate with its nearest neighbours to the NW, W and SW, and the NE, E and SE directions respectively. Control circuits in neighbouring cells communicate via single-bit control lines indicated in Figure 17.4 as REQuest (REQ) and AVAILability (AVAIL) signals.

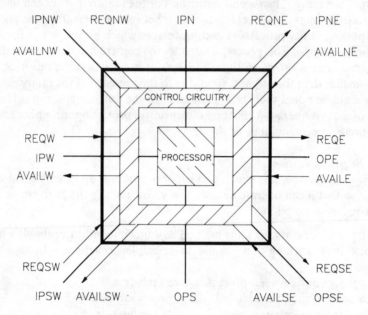

Fig. 17.4 *Schematic of cell suitable for one-dimensional WINNER algorithm.*

17.3.2 Control circuitry

The function of the control circuitry in each cell is to decide how the cell should be connected to its neighbours in adjacent columns. In addition it must decide whether or not it should act as a bypass in the north-south direction. These decisions must be made on the basis of the following information, which is the only information available to the control circuitry:

1. a knowledge of whether the processor in its cell is functional or faulty
2. REQuest inputs from its NW, W and SW neighbours
3. AVAILability inputs from its NE, E and SE neighbours

Using this information the control circuitry must perform the following functions:

1. generate REQuest and AVAILability signals and route them to neighbouring cells
2. select data inputs from the appropriate neighbour and apply them to the processor
3. bypass the processor in North-South direction if the processor is faulty or unused

Convention for REQuest and AVAILability signals: The following convention for REQuest and AVAILability signals will be used:

REQuest and AVAILability signals are *active* when at a logic 1 level and *passive* when at a logic 0 level.

Based on this convention, the phrases *to send an AVAILability signal* and *to send a REQuest* imply that the signals being sent are TRUE.

If a cell A sends an AVAILability signal to another cell B, it means that cell A contains a processor which is AVAILable for connection if REQuested by cell B. If a cell A sends a REQuest signal to some other cell B, it means that cell A wishes to set up a communication channel between its processor and the processor in B. Cell A can only send a REQuest to cell B if B is sending and AVAILability signal to cell A. If such a communication channel becomes set up then A is said to have been *connected* to B and the incoming data signals to cell B from cell A are directed to the processor in cell B. The manner in which these signal are generated by a cell forms the heart of the algorithm and is described in detail in the following sections.

Generation of AVAILability signals: A cell generates AVAILability output signals according to the following rules:

1. A cell can only output a TRUE AVAILability signal if it contains a processor which is fault-free (i.e. the self-test shows it to be functional), and at least one TRUE AVAILability signal is being received from its NE, E or SE neighbours.
2. If 1 is satisfied, then the priority system given in table 17.1 operates to decide in which directions to send TRUE availability signals depending upon the incoming REQuest signals

From the bottom line of table 17.1 we can see that a cell receiving no REQuests will output a TRUE AVAILability signal to each of its three left hand neighbours. This allows any of the neighbours to make a REQuest if it wishes to do so at a later time. From the first three lines of the table we see that once a REQuest has been received, the cell may or may not remain AVAILable to the other neighbours depending on the priority of the REQuest. A REQuest from

Table 17.1: *Generation of AVAILability output signals.*

REQuest Inputs			AVAILability Outputs		
REQNW	REQW	REQSW	AVAILNW	AVAILW	AVAILSW
TRUE	*X*	*X*	TRUE	FALSE	FALSE
FALSE	TRUE	*X*	TRUE	TRUE	FALSE
FALSE	FALSE	TRUE	TRUE	TRUE	TRUE
FALSE	FALSE	FALSE	TRUE	TRUE	TRUE

X = TRUE or FALSE

the NW neighbour has the highest priority, with the W and SW neighbours having successively lower priorities.

The scheme allows a priority of connections to be established so that a REQuest from the NW has highest priority, and REQuests from the W and SW have successively lower priorities. Such a scheme is required in a closed system to ensure that a stable solution is reached. The scheme causes cells to output FALSE AVAILability signals to neighbouring cells if they have no chance of obtaining a connection. This occurs for example when a higher priority connection has already been established.

Rule 1 above gives the cell a global look-ahead capability even though each cell is capable only of local communication. This enables clustered faults to be avoided in the following way. Information is passed between cells from east to west about the availability of other cells. This allows a cell A to prohibit another cell from connecting to it if A either contains a faulty processor or would be part of a dead-end route – i.e., a route that would not be able to be completed due to some blockage later. Such a dead end route could occur, for example, if three vertically adjacent processors were faulty. In this case a functional processor to the left of the centre faulty processor would find that all of its possible connections to neighbours are unavailable and output FALSE AVAILability signals. The functional processor would then declare itself to be unavailable. The scheme allows information about blockages to be transmitted from right to left to all the relevant processors, which then decide upon some appropriate avoiding action. These features will be illustrated in 17.3.5.

Generation of REQuest signals: Request signals are output from a cell according to a different set of rules:

1. A cell can only output a TRUE REQuest signal if its processor is fault-free and at least one REQuest has been received from one of its NW, W or SW neighbours

2. If 1 is satisfied then the cell outputs a single TRUE REQuest value to one of its NE, E and SE neighbours depending upon the incoming AVAILability signals according to the priority given in table 17.2

These rules ensure that only one REQuest signal is output from any cell,

Table 17.2: *Generation of REQuest output signals.*

AVAILability Inputs			REQuest Outputs		
AVAILNE	AVAILE	AVAILSE	REQNE	REQE	REQSE
TRUE	*X*	*X*	TRUE	FALSE	FALSE
FALSE	TRUE	*X*	FALSE	TRUE	FALSE
FALSE	FALSE	TRUE	FALSE	FALSE	TRUE
FALSE	FALSE	FALSE	FALSE	FALSE	FALSE

X = TRUE or FALSE

which in turn ensures that a cell can never accidentally become connected to more than one neighbouring cell in any column.

17.3.3 Array boundary condition

When the WINNER cells are connected in an array, the inputs around the edge of the array are not connected to other cells and must be defined explicitly. The inputs to the REQuest and AVAILability lines are defined as follows.

Boundary AVAILability inputs:

- Set NE and SE boundary REQuest inputs to FALSE
    - Set E boundary AVAILability inputs to TRUE

Boundary REQuest inputs:

- Set NW and SW boundary REQuest inputs to FALSE
    - Set each W boundary REQuest inputs to TRUE if corresponding W AVAILability output from the array is TRUE; otherwise set to FALSE

17.3.4 Interaction of REQuest and AVAILability signals

The AVAILability and REQuest signals together provide the cells with all the information they need about their surroundings in order to be able to form functional rows of interconnected processors. The priority system for sending and receiving REQuest and AVAILability signals ensures that stable functional rows are established from W to E and from N to S starting in the top left-hand corner of the array. The priority system also ensures that each row formed is as close as possible to the northern edge of the array, thus maximising the number of rows generated.

The one dimensional WINNER algorithm operates automatically when the array is switched on and is a totally asynchronous technique. All cells continuously make decisions based on the information available to them. This information will be changing as the organisation of the array gradually evolves to its stable state. Therefore, in the early stages of self-organisation the array may be highly dynamic with cells forming and relinquishing connections to other cells as a result of being overidden by higher priority decisions which have been made at other localities and have rippled through the array. Connections may in fact experience a number of iterations of this type but due to the priorities for generation of the REQuest and AVAILability signals, the array will always settle into a self-consistent, stable state.

The array can be visualised as having two levels of hierarchy. One level comprises the underlying asynchronous network of control circuitry, which is capable of establishing communication channels between appropriate neighbours to generate a functionally orthogonal array as described. The second level is the array of processors containing a number of faulty elements, which can be considered to be overlaid on the array of control circuitry. The control network then forms connections between proessors as appropriate.

17.3.5 Serial description of WINNER operation

Although the interactions of REQuest and AVAILability signals between the cells of the array occur simultaneously, it is helpful from the point of view of understanding the operation of the algortithm to consider the process as a *sequence* of distinct events as follows.

We assume that initally no REQuest or AVAILability signals are present in the array other than the fixed boundary input values. From this starting point, no REQuest signals can be generated by any cells until at least one AVAILability signal has reached the left hand side of the array; this is the first stage of the configuration process. The AVAILability signals from each cell to its neighbours are generated starting from the right-hand column of the array and working column by column across the array. For each column, the three AVAILability outputs of each cell in the column are set to TRUE only if the cell has at least one TRUE incoming AVAILability signal. This procedure has the effect of flushing out the dead end routes from the array. Any dead end route will result in all the cells on that route outputting FALSE AVAILability signals.

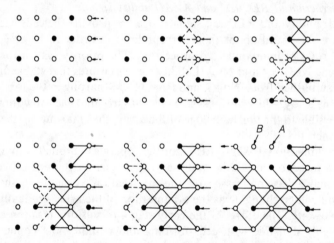

Fig. 17.5 *Step-by-step generation of AVAILability signals showing avoidance of dead-end routes.*

This is illustrated in Figure 17.5, which shows the column by column movement of AVAILability signals across the array. A continuous line between two cells indicates the presence of a TRUE AVAILability output, while dashed lines represent the TRUE AVAILability output signals currently being generated.

The faulty cells (marked by a cross) are obviously un-AVAILable; however, in addition, some of the functional elements are shown as being un-AVAILable.

The cell labelled A is un-AVAILable since all of its right-hand neighbours are faulty, while the cell labelled B is un-AVAILable because although not all of its right-hand neighbours are faulty, none of them is AVAILable.

In the second stage of the process REQuest signals are generated starting from the left hand column and working from left to right across the array. This is illustrated in Figure 17.6 in which we consider an input REQuest being applied only to the uppermost cell in the left hand column which is outputting a TRUE AVAILability signal. Further REQuest signals are then generated in subsequent columns by following a path of TRUE AVAILability signals according to the rules given in Table 17.2. In other words, REQuest signals follow a path as close to the top edge of the array as possible. In this way a complete functional row can be constructed without backtracking since a cell outputting a TRUE AVILability signal indicates that a continous path exists between the left and right hand sides of the array.

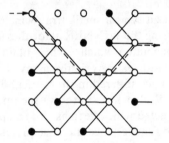

Fig. 17.6 *Generation of REQuest signals for first functional row.*

Subsequent functional rows are generated by alternate cycles of AVAILability and REQuest generation until all possible functional rows have been formed. This procedure ensures that any processors which have become *un*available as a result of the presence the first functional row will output the appropriate modified AVAILability signals.

17.3.6 Five-neighbour WINNER algorithm
In the foregoing description of the WINNER algorithm only nearest neighbour interconnections between adjacent columns were permitted. This resulted in a cell which could communicate with three neighbouring cells in both the column to the left and the right. However, this restriction is not due to any fundamental limiting property of WINNER, and the algorithm can be extended in a straightforward manner to a scheme which allows a cell to communicate with any of five neighbouring cells in the column to the left or right.

The main advantage of a five-neighbour interconnection scheme is that configuration performance is improved. This is because of the introduction of the longer range communication, which allows some cells to be used in the configured array that would otherwise have been omitted from the array. The truth tables for the control logic can be extended in a straightforward manner to handle the extra inputs and outputs and will not be further described.

17.4 Algorithm 2: WINNER in two dimensions

Algorithm 1 makes the basic assumption that it is always possible to bypass faulty cells in the vertical direction. In many cases this may be a valid assumption and the technique could be used successfully in many applications. However, when we started this research our basic philosophy was that the configured array should completely avoid all faulty processors. In this section we will show how the one-dimensional WINNER algorithm can be extended to encompass this philosophy by applying it to both the rows and the columns simultaneously, so that faulty processors are completely avoided. To see how this is done the reader is referred to Figure 17.7.

Figure 17.7(a) illustrates a typical small array which has been configured in the horizontal direction using the WINNER algorithm in one dimension. Functional rows have been generated which each contain a number of working processors equal to the width of the original array. Figure 17.7(b) shows the same array which has been configured in the vertical direction using the same one-dimensional WINNER algorithm but this time operating on the columns. Here, columns are constructed which keep as close as possible to the left-hand side of the array and each functional column is equal in length to the height of the original array.

Since the configurations generated by the algorithm consist of full width rows and full height columns, one might suppose that the superposition of the rows and the columns would generate an array of functional processors at the points of intersection of each row with each column, and that the processors within the array region would have orthogonal interconnections. The simple superposition of separately configured rows and columns is shown in Figure 17.7(c), where the processors forming the composite array are indicated in black. Some processors have either horizontal or vertical connections but not both. These are unused processors and are controlled to act as bypasses in the direction in which they have connections. The size of the functional array is determined by the number of functional rows and columns generated. If p is the number of functional rows and q the number of functional columns, then simple superposition should generate a functional array of dimensions $p \times q$. A cell suitable for use with this algorithm requires communication and control signal paths for both the horizontal and vertical directions and is illustrated schematically in Figure 17.8.

At first sight this simple superposition process appears to be quite straightforward. However, there are in fact two types of undesirable condition which can occur when the rows and columns are superimposed in such a simple manner. These have been called *double site* and *crossover* conditions and must be handled within the algorithm if a correctly configured array is to be produced for all fault

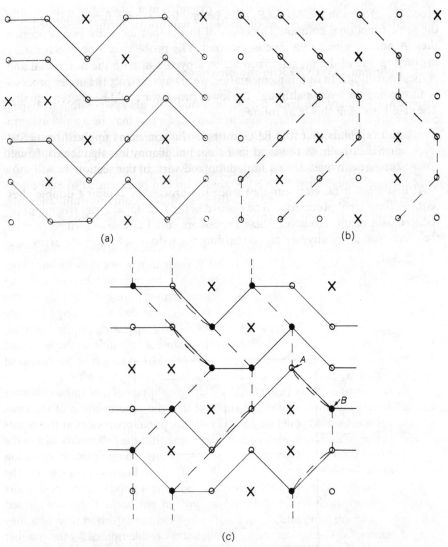

Fig. 17.7 *The WINNER algorithm applied in two dimensions: (a) Configured rows, (b) Configured columns, (c) Superposition of the functional rows and columns to create a functional array.*

distributions. The example illustrated here contains several double sites. As we shall see, a simple technique has been developed which results in an algorithm capable of configuring a functional array from any fault distribution.

17.4.1 Double site condition
Referring to the small array configured by simple superposition of functional rows and columns shown in Figure 17.9(a) we see that there are pairs of

processors, for example A and B, both belong to the same functional row and the same functional column. The effect of this is that there are two processor sites, A and B, where only one is required. The problem can be overcome by instructing one of the processors to act as a bypass in both the horizontal and vertical directions. In our implementation we always instruct the upper processor to become the bypass although the lower processor could be chosen equally. The rule for doing this is as follows:

> If a cell finds that it is REQuesting to be connected to a cell to its SW or SE in both its row and its column configuration algorithms, it will become a bypass for its horizontal and vertical connections.

At first sight it may appear that since the process of avoiding double sites requires functional processors to be discarded the functional array size might be reduced as a result. However, these processors could not have formed part of the functional array anyway and discarding them does not affect the array size of $p \times q$.

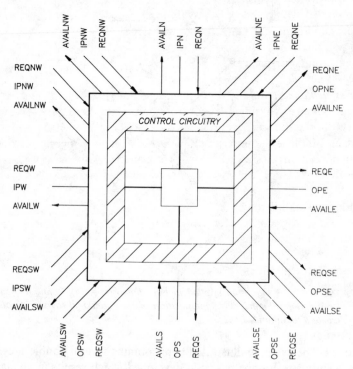

Fig. 17.8 *Schematic of cell suitable for two-dimensional WINNER algorithm.*

17.4.2 Crossovers

Crossovers occur when a row and a column intersect each other at a point other than a processor site. Crossovers can occur in two distinct ways as shown in

Figures 17.9(*b*) and (*c*). Unlike with the double-site condition, which can be overcome without altering the configured rows and columns, the solution to the crossover condition requires either the functional row or functional column containing the crossover to be physically altered so that the superposition crossover does not occur. Alteration of the row or column may of course produce perturbations that propagate throughout the array until a new stable configuration is achieved.

The two fundamental modes in which crossovers can occur are shown in Figure 17.9(*b*) and (*c*). Other crossovers can occur, such as those shown in Figure 17.9(*d*) and (*e*) but these simply superpositions of the fundamental modes. In mode 1 of Figure 17.9(*b*) the crossover can be avoided by making cell 3 unavailable to cell 2 in the presence of the link between cells 1 and 4 (link 1–4). Alternatively, cell 4 could be made unavailable to cell 1 in the presence of link 2–3. In a similar manner, the mode 2 crossovers in Figure 17.9(*c*) can be avoided by making cells 3 or 4 unavailable in the presence of links 1–4 or 2–3 respectively.

We have proposed two solutions to the crossover problem, either of which can be embodied within the final two dimensional WINNER algorithm. The first requires an additional single-bit control signal to pass between cells in the E-W and N-S directions while the second requires a change only to the logic of the control circuitry in each cell.

Crossover avoidance using extra control communication: This technique requires the use of an extra, single-bit control signal which must pass between cells in both the horizontal and vertical directions. This allows the rows to be generated as in the one-dimensional WINNER algorithm but restricts the generation of columns to sites which will not cause crossovers. As we have seen from Figure 17.9(*b*), the crossover could be avoided if cell 3 was made un-AVAILable to cell 2, and in Figure 17.9(*c*) the crossover could be avoided if cell 4 was made unavailable to cell 1. The first of these can be achieved by using a single extra bit which propagates between cells from north to south. The extra bit indicates whether or not the cell which generated it is outputting a row REQuest in the SE direction and if so causes the column AVAILNE signal in the cell below to be inhibited. The second crossover mode may be avoided in a similar manner by passing an extra bit from west to east, indicating whether a cell has output a row REQuest to the NE, and if so inhibiting the column AVAILNW in the cell to its right. The cost of this technique is two extra inputs and outputs plus two (A AND NOT B) logic functions to perform the inhibitory action.

Crossover avoidance by alteration to control circuitry: The ideal solution to the crossover problem would be one involving a simple alteration to the control circuitry in each cell without requiring any extra connections between cells, since this is likely to introduce the smallest overhead of area into the algorithm. An approach incorporating these concepts has been developed and is now described.

To avoid mode 1 crossovers we wish to make cell 3 unavailable to cell 2 in the presence of link 1–4. This can be achieved without using extra control bits by noting that if link 1–4 exists, then because this is the highest priority direction for the row generation circuitry in cell 4, cell 4 will output a FALSE AVAIL-

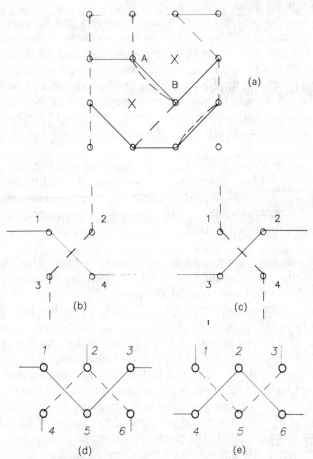

Fig. 17.9 *Double-site and crossover modes which can occur with simple superposition of functional rows and columns: (a) double-site condition; (b) fundamental crossover mode 1; (c) fundamental crossover mode 2, (d) composite crossover mode; (e) composite crossover mode.*

ability signal to cell 3. If the 1–4 link does not exist, then cell 4 outputs a TRUE AVAILability signal to cell 3. This means that the 1–4 link can be detected by cell 3 by the value of the AVAILability signal coming from cell 4. The AVAILability signal can therefore be used to modify the AVAILability of cell 3 in the

NE direction of the column generation circuitry, i.e., to cell 2. This technique can be implemented by ANDing the NE column AVAILability output signal of each cell with the incoming eastern row-AVAILability signal. This gives the row link priority over the column link, which must then find an alternative route.

In a similar manner, mode 2 crossovers can be avoided. In this case, a FALSE column AVAILability output from cell 4 to cell 2 indicates that the link 1–4 is present. This information can then be used in cell 2 to inhibit (with a single AND gate) its row AVAILability to cell 3. An alternative row route will then have to be found.

17.5 Performance of the WINNER algorithm

The WINNER algorithm applied to one dimension of the array has been extensively simulated to evaluate its ability to configure arrays with different fault distributions. In this section we present some simulation results relating to the ability of the technique to configure a 10 × 10 functional array from a larger array containing faults. Simulations have been carried out for a variety of initial array sizes and processing element yields. For each value of array size and processor yield, a large number of random fault distributions were applied to the array and the number of arrays capable of being configured into at least a 10 × 10 functional array were counted. From this information, the percentage overhead of processing elements required to achieve, on average, a 50% array yield for the 10 × 10 target array was calculated for a range of values of processor yield.

The results of the simulations are presented in Figure 17.10 together with corresponding results obtained by simulation of some other published configuration strategies. The dashed curve represents the theoretical upper bound on configuration performance and would be very difficult and costly in terms of hardware to achieve in practice. It can be seen that the WINNER performance is comparable with the other techniques when the processor yield is greater than 80%. At processor yields of less than 80% the techniques of Moore and Mahat (1985) and Sami and Stefanelli (1983) have progressively better performance than WINNER. The performance of both triple modular redundancy and the scheme of Hedlund and Snyder (1982) are significantly poorer than the other techniques.

17.6 Hardware requirements of WINNER

The circuitry required to implement control logic is illustrated in Figure 17.11. The circuit has been drawn for a processing element having a single connection in the N, S, E and W directions and with data flowing from N to S and from

E to W. Circuitry to handle extra data lines or bi-directional data-flow can be derived in a straightforward manner.

17.7 Advantages of the WINNER technique

The WINNER self-organising approach to the configuration of two-dimensional processor arrays is quite different from other published techniques and has several distinct advantages as follows:

1. The array can configure itself automatically without the need for external assistance
2. The self-organising algorithm is fully convergent and cannot become unstable

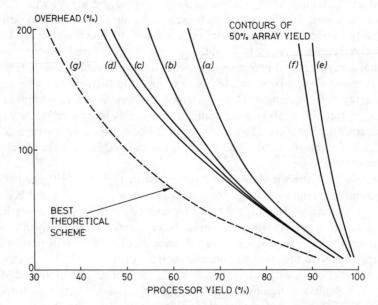

Fig. 17.10 *Comparison of the performance of various configuration schemes. Curves as follows: (a) 3-neighbour WINNER; (b) 5-neighbour WINNER; (c) Moore and Mahat (1985); (d) Sami and Stefanelli (1983); (e) triple modular redundancy; (f) Hedlund and Snyder (1982); (g) theoretical upper bound.*

3. The control circuitry associated with each cell in the array is simple, about 20 gates, resulting in a low overhead
4. Good utilisation of functional processors particularly when processor yield is greater than 80%
5. No global control lines are required
6. The array is potentially capable of being reconfigured in the event of an

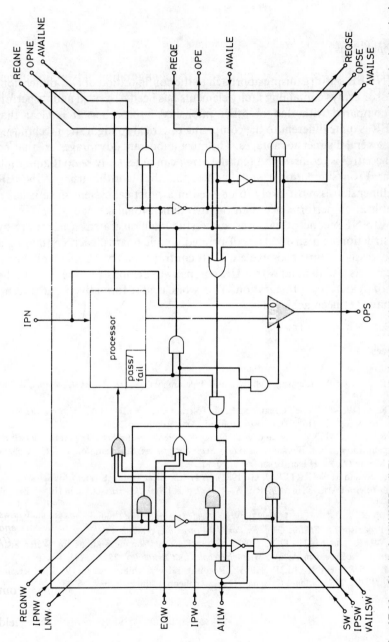

Fig. 17.11 *Circuit of a WINNER cell. The gates shown shaded are used for routing data; the remainder represent the decision logic required to configure the array.*

in-service failure and could therefore be useful for remotely sited equipment, or in equipment requiring a very fast repair time

17.8 Conclusions

The WINNER self-organising approach to the configuration of two-dimensional processor arrays containing faulty elements has been described and its performance compared with that of other published approaches. It is clear that WINNER is quite different to the competing electrical configuration techniques and has several distinct advantages. The most important advantages are that no external control is required and that the array can potentially *re*-configure itself in-service. From simulation it would appear that the algorithm using WINNER in one dimension is most likely to be used in a practical system since it has a better utilisation performance than the two-dimensional scheme.

The WINNER concept has been successfully demonstrated in hardware by the construction of a small array of printed circuit boards, each containing a single processing element complete with its control circuitry. Further collaborative research is also in hand with a UK company to demonstrate the ideas in the context of Wafer Scale Integration. This work is at a formative stage and no results have yet been achieved.

References

EVANS, R. A. (1985a): Processing Cell for Fault Tolerant Arrays, *UK Patent Application No.* 85 02186

EVANS, R. A. (1985/6): 'A Self Organising Fault Tolerant, 2-Dimensional Array', *Proc. VLSI-85, Tokyo*, Japan, (ed., E. Hoebst), (North-Holland, Amsterdam) pp. 239–48

EVANS, R. A. McCANNY, J. V. and WOOD, K. W. (1985): 'Wafer Scale Integration Based on Self-Organisation', *Proc. Workship on Wafer Scale Integration*, Southampton, eds. C. Jesshope and W Moore (Adam Hilger, Bristol), pp. 101–12

EVANS, R. A. and McWHIRTER, J. G. (1987): 'A Hierarchical Testing Strategy for Self-organising Fault-tolerant Arrays', in MOORE, W. R. *et al.* (eds.) *Systolic Arrays*, (Adam Hilger, Bristol) pp. 229–38

HEDLUND, K. S. and Snyder, L. (1982): 'Wafer Scale Integration of Configurable Highly Parallel (CHiP) Processor', *IEEE Int. Conf. on Parallel Processing*, pp. 262–4

MOORE, W. R. and MAHAT, R. (1985): 'Fault Tolerant Communications for Wafer Scale Integration of a Processor Array', *Microelectronics and Reliability*, **25**, pp. 291–4

SAMI, M. and STEFANELLI, R. (1983): 'Reconfigurable Architectures for VLSI Processing Arrays', *IFIPS 1983 National Computer Conf.*, Anaheim, California, pp. 565–77

Index